For further volumes:
http://www.springer.com/series/4205

Guanghui Wang • Q.M. Jonathan Wu

Guide to Three Dimensional Structure and Motion Factorization

Springer

Guanghui Wang
Department of Systems Design Engineering
University of Waterloo
200 University Avenue West
Waterloo
Ontario N2L 3G1
Canada
guanghui.wang@uwaterloo.ca

Q.M. Jonathan Wu
Dept. Electrical & Computer Engineering
University of Windsor
401 Sunset Avenue
Windsor
Ontario N9B 3P4
Canada
jwu@uwindsor.ca

Series Editor
Professor Sameer Singh, PhD
Research School of Informatics
Loughborough University
Loughborough
UK

ISSN 1617-7916
ISBN 978-0-85729-045-8 e-ISBN 978-0-85729-046-5
DOI 10.1007/978-0-85729-046-5
Springer London Dordrecht Heidelberg New York

British Library Cataloguing in Publication Data
A catalogue record for this book is available from the British Library

Library of Congress Control Number: 2010936442

Cover design: VTEX, Vilnius

Printed on acid-free paper

Springer is part of Springer Science+Business Media (www.springer.com)

Preface

Structure and motion recovery refers to the process of extracting three dimensional structure of the scene as well as camera motions by analyzing an image sequence. This is an important theme in computer vision, and great progress has been made both in theory and in practice during the last two decades. Successful applications include robot navigation, augmented reality, industrial inspection, medical image analysis, digital entertainment, and many more.

The book focuses on Euclidean structure and motion recovery, particularly on factorization based algorithms. The factorization method is based on bilinear formulation that decomposes image measurements directly into structure and motion components. Recent studies have extended the algorithm to recover structure of both nonrigid and dynamic objects, and experimental results have established its efficiency and potency. The book provides a comprehensive overview and an in-depth study of rigid and nonrigid factorization techniques. Our recently developed algorithm along with some latest results are presented in the book.

The book is suitable for graduate students, researchers, and industrial practitioners. Some background on projective geometry and matrix computation would greatly assist in understanding the fundamental ideas presented in the book. However, no profound knowledge of computer vision is required. The main contents of this book are as follows.

The first three chapters cover camera imaging geometry and their applications. Chapter 1 introduces some basic concepts and principles of imaging geometry, such as single and two-view geometry, image metrology, and three dimensional reconstruction. Chapter 2 reviews affine approximation model and presents a quasi-perspective projection model as a trade-off between simplicity and accuracy of affine and perspective projection models respectively. Chapter 3 investigates geometrical properties of quasi-perspective projection model, quasi-perspective projection matrix, fundamental matrix, plane induced homography, and quasi-perspective reconstruction.

Chapter 4 introduces basic principles of various affine and perspective factorization algorithms for structure and motion recovery of rigid, nonrigid, articulated, and multiple moving objects.

Chapters 5 and 6 deal with perspective extension of affine factorization algorithm. Chapter 5 describes two techniques to improve the performance of rigid perspective factorization, one is a hybrid approach for projective depth estimation based on projective reconstruction, and the other is a Kruppa constraints based self-calibration scheme for a more universal camera model. Chapter 6 presents two algorithms for perspective factorization of nonrigid objects, the first one refines the depth scales using linear recursive estimation, and the second one is based on nonlinear optimization via minimization of perspective reprojection residuals.

Chapter 7 describes an alternative rotation constrained power factorization algorithm that efficiently integrates orthonormality and replicated block structure of the motion matrix into an iterative scheme.

Chapter 8 introduces a deformation weight constraint for nonrigid factorization and proves the invariability between structure and shape bases under Euclidean and affine transformations. Hence, the metric structure can be recovered using a stratification approach.

Chapter 9 applies the quasi-perspective projection model to structure and motion recovery and establishes a factorization framework for rigid and nonrigid objects under quasi-perspective assumption. Computational details are elaborated in the chapter and an extended Cholesky decomposition is proposed to recover the rotation part of the Euclidean upgrading matrix.

The appendix presents some mathematical facts that are widely used in the book, including projective geometry, matrix decomposition, least squares, and nonlinear estimation techniques.

Acknowledgements We would like to thank Dr. Zhanyi Hu, Dr. Hung-Tat Tsui, Dr. John Zelek, and Professor Fuchao Wu for their constant advice and support. We thank our colleagues and friends: Jian Yao, Yifeng Jiang, Jun Xie, and Wei Zhang for inspiring discussion. We also acknowledge Abdul Adeel Mohammed for proofreading the text and offering insightful suggestions.

The core contents of the book are based on our publications, we benefited greatly from comments and suggestions by many anonymous reviewers. This research was supported in part by Natural Sciences and Engineering Research Council of Canada and the National Natural Science Foundation of China.

The book could not have been completed without the excellent support and professional instructions of Simon Rees and Wayne Wheeler at Springer-Verlag, London. We also wish to show our deep gratitude to everyone who helped throughout the preparation of this book.

Finally, we would like to thank our families for their consistent support.

Waterloo, Canada — Guanghui Wang
Windsor, Canada — Jonathan Wu

Contents

Acronyms

2D	Two-dimensional
3D	Three-dimensional
AAM	Active appearance model
CCD	Charge-coupled device
CPF	Constrained power factorization
DIAC	Dual image of the absolute conic
DOF	Degrees of freedom
EM	Expectation-maximization algorithm
IAC	Image of the absolute conic
KLT	Kanade-Lucas-Tomasi feature tracker
LM	Levenberg-Marquardt algorithm
MM	Morphable model
PF	Power factorization
QR	QR factorization
RANSAC	Random sample consensus
RCPF	Rotation constrained power factorization
RQ	RQ factorization
SFM	Structure from motion
SIFT	Scale-invariant feature transform
STD	Standard deviation
SVD	Singular value decomposition
VRML	Virtual reality modeling language

Introduction to 3D Computer Vision 1

Abstract This chapter introduces some basic concepts and ideas of computer vision, such as imaging geometry of cameras, single view geometry, and two-view geometry. In particular, the chapter presents two practical examples. One is on single view metrology, calibration, and reconstruction; the other is a hybrid method for reconstruction of structured scenes from two uncalibrated images.

Mathematics is the tool specially suited for dealing with abstract concepts of any kind and there is no limit to its power in this field.

Paul Adrien Maurice Dirac (1902–1984)

1.1 Introduction

Making a computer see and understand the world is the main task in computer vision. During the past two to three decades, great progress has been made in both theory and applications. Computer vision based systems and products have been widely used in industry, medical image processing, military, autonomous vehicles, digital entertainment, and many aspects of our daily life.

In this chapter, we will briefly review some basic knowledge and principles of computer vision, such as camera imaging geometry, calibration, single view and two-view geometry, 3D reconstruction, etc. These ideas are widely used in literature and the later part of this book. For a better understanding, the reader is expected to be familiar with these basic ideas and some notations of projective geometry as outlined in Appendix.

We start with the introduction of imaging geometry and camera models in Sect. 1.2, then present the properties of forward and back-projection of some geometric entities by the camera. Followed by a practical application on single view based measurement and 3D reconstruction in Sect. 1.3. The two view geometry and stereo vision are presented in Sect. 1.4, which include epipolar geometry, fundamental matrix, and stereo vision. Finally, we present a practical method for two view reconstruction of structured scenes based on point and line features.

G. Wang, Q.M.J. Wu, *Guide to Three Dimensional Structure and Motion Factorization*, Advances in Pattern Recognition,
DOI 10.1007/978-0-85729-046-5_1,

1.2 Imaging Geometry and Camera Models

We are familiar with Euclidean geometry since it is widely applied to describe the world around us. However, it is insufficient when we consider the mapping from space to an image as some Euclidean properties are not preserved during the process. Projective geometry is introduced to model the imaging process, which allows a more general class of transformations than just translations and rotations in Euclidean geometry. Please refer to Appendix A or [8, 11] for some basic concept and properties of projective geometry. In this section, we will present a brief introduction of camera models and imaging geometry.

1.2.1 Camera Models

In projective geometry, a point is normally denoted in homogeneous form. Suppose $\bar{\mathbf{x}} = [u, v]^T$ is a 2D image point and $\bar{\mathbf{X}} = [x, y, z]^T$ is a 3D space point, then the corresponding homogeneous forms are expressed as $\mathbf{x} = [\bar{\mathbf{x}}, 1]^T$ and $\mathbf{X} = [\bar{\mathbf{X}}, 1]^T$ respectively.

The principal cameras in computer vision are based on pinhole camera model that projects features in 3D world to features in 2D image. When homogeneous coordinates are employed, the mapping from space to image can be expressed linearly as follows regardless of the choice of coordinate frames.

$$\lambda \begin{bmatrix} u \\ v \\ 1 \end{bmatrix} = \underbrace{\begin{bmatrix} p_{11} & p_{12} & p_{13} & p_{14} \\ p_{21} & p_{22} & p_{23} & p_{24} \\ p_{31} & p_{32} & p_{33} & p_{34} \end{bmatrix}}_{\mathbf{P}} \begin{bmatrix} x \\ y \\ z \\ 1 \end{bmatrix} \tag{1.1}$$

where λ is a scale factor, usually called the projective depth; $\mathbf{P}$ is a 3×4 matrix known as perspective projection matrix or camera matrix.

The imaging process of a pinhole camera is shown in Fig. 1.1. In the simplest case where the optical center of the camera is placed on the origin of world coordinate frame and the focal length is equal to 1, the camera matrix $\mathbf{P}$ is simplified to

$$\mathbf{P}_0 = \begin{bmatrix} 1 & 0 & 0 & 0 \\ 0 & 1 & 0 & 0 \\ 0 & 0 & 1 & 0 \end{bmatrix} = [\mathbf{I}_3 | \mathbf{0}] \tag{1.2}$$

where $\mathbf{I}_3$ denotes a 3×3 identity matrix. The matrix $\mathbf{P}_0$ is referred as a canonical form of the camera matrix.

In a general case that the world system is different from the camera system, the relationship between the two systems can be defined by a 3D rotation $\mathbf{R}$ followed by a translation $\mathbf{T}$, which is equivalent to multiplying matrix $\mathbf{P}_0$ to the right by a 4×4 transformation

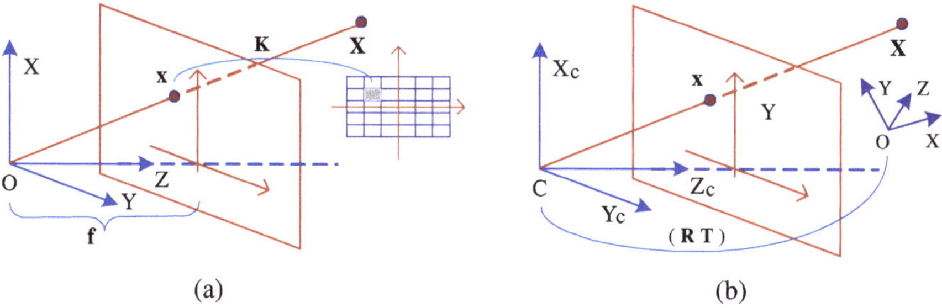

Fig. 1.1 Perspective camera geometry. (**a**) A special case where the optical center of the camera coincides with the origin of the world system; (**b**) A general case with arbitrary world and camera systems

matrix

$$\mathbf{A} = \begin{bmatrix} \mathbf{R} & \mathbf{T} \\ \mathbf{0}^T & 1 \end{bmatrix}, \quad \mathbf{R} = \begin{bmatrix} \mathbf{r}_1^T \\ \mathbf{r}_2^T \\ \mathbf{r}_3^T \end{bmatrix}, \quad \mathbf{T} = \begin{bmatrix} t_x \\ t_y \\ t_z \end{bmatrix} \tag{1.3}$$

where $\mathbf{R}$ is a 3×3 rotation matrix, which has 3 degrees of freedom (DOF) and $\mathbf{R}\mathbf{R}^T = \mathbf{I}_3$; $\mathbf{r}_i^T$ stands for the ith row of the rotation matrix. $\mathbf{T}$ is a 3D translation vector. The matrix $\mathbf{A}$ describes the pose of the camera, which is known as the extrinsic parameters of the camera. It has 6 independent parameters, 3 for the rotation and the remaining 3 for the translation.

For an actual camera, the image coordinates do not correspond to the physical coordinates in the retinal plane. The relationship between the two systems can be modelled by a 3×3 upper triangular matrix $\mathbf{K}$ which depends on the intrinsic parameters of the camera, such as the size and shape of the pixels and the position of the imaging device. Change of coordinates in the image plane is equivalent to multiplying matrix $\mathbf{P}_0$ to the left by the matrix $\mathbf{K}$.

$$\mathbf{K} = \begin{bmatrix} f/s_u & f/s_u \cot\theta & u_0 \\ & f/s_v & v_0 \\ & & 1 \end{bmatrix} = \begin{bmatrix} f_u & \varsigma & u_0 \\ & f_v & v_0 \\ & & 1 \end{bmatrix} \tag{1.4}$$

where f represents the focal length of the camera; s_u and s_v stand for the width and height of the pixel footprint; $[u_0, v_0]^T$ is the coordinates of the camera's principal point, i.e. the intersection of the optical axis and the retinal plane; θ is the angle between the two image axes.

The matrix $\mathbf{K}$ is known as the camera calibration matrix. We define $\varsigma = f/s_u \cot\theta$ as the skew parameter of the camera, and $\kappa = s_u/s_v$ as the camera aspect ratio. For some precise industrial CCD cameras, it is safe to assume unit aspect ratio and orthogonal image axes, i.e. $\kappa \to 1$, $\theta \to \pi/2$. Thus we have $f_u = f_v$ and $\varsigma_i = 0$. The principal point is usually at the center of the image, i.e. $u_0 = v_0 = 0$. Then the camera is simplified to have only one intrinsic parameter.

Taking account of the coordinates transformation (1.3) and (1.4), a general perspective projection matrix becomes

$$\mathbf{P} = \mathbf{K}\mathbf{P}_0\mathbf{D} = \mathbf{K}[\mathbf{R}|\mathbf{T}] \tag{1.5}$$

and the imaging process of perspective projection can be written as

$$\lambda\mathbf{x} = \mathbf{P}\mathbf{X} = \mathbf{K}[\mathbf{R}|\mathbf{T}]\mathbf{X} \tag{1.6}$$

A general projection matrix has 12 parameters defined up to an arbitrary scale, thus it has only 11 degrees of freedom (6 for extrinsic parameter **R** and **T**, and the rest 5 for intrinsic parameters **K**). From (1.6) we observe that each space to image point correspondence provides two constraints on **P**. Thus the projection matrix can be recovered from a minimum $5\frac{1}{2}$ such correspondences at a general position.

We usually denote the camera projection matrix in the following form

$$\mathbf{P} = [\mathbf{M}, \mathbf{p}_4] = [\mathbf{p}_1, \mathbf{p}_2, \mathbf{p}_3, \mathbf{p}_4] = \begin{bmatrix} \mathbf{p}_1^T \\ \mathbf{p}_2^T \\ \mathbf{p}_3^T \end{bmatrix} \tag{1.7}$$

where $\mathbf{p}_i^T$ denotes the ith row of **P**, and matrix **M** stands for the first three columns of **P**. The camera matrix has the following geometric properties and meanings.

Result 1.1 *The projection matrix is of rank-3 and it has 1-dimensional right null-space* **C** *that satisfies* $\mathbf{PC} = 0$. *It can be verified that* **C** *is the homogeneous representation of the camera center. If* **M** *is not singular,*

$$\mathbf{C} = \begin{bmatrix} -\mathbf{M}^{-1}\mathbf{p}_4 \\ 1 \end{bmatrix}$$

Otherwise, if **M** *is singular and* **d** *is the right null vector of* **M**, *we have*

$$\mathbf{C} = \begin{bmatrix} \mathbf{d} \\ 0 \end{bmatrix}$$

which means the camera center is located at infinity. We call it camera at infinity.

Result 1.2 *The principal plane is the plane passing through camera center and parallel to the image plane, which is equal to the last row* $\mathbf{p}^3$ *of the camera matrix. The principal axis is the line passing through the camera center and perpendicular to the principal plane. The principal axis vector* $\mathbf{v} = |\mathbf{M}|\mathbf{m}^3$ *is in the direction of the principal vector and directed towards the front of the camera. The intersection of the principal axis and the image plane is called principal point, which can be computed from* $\mathbf{x}_0 = \mathbf{M}\mathbf{m}^3$, *where* $\mathbf{m}^3$ *denotes the third row of* **M**.

Result 1.3 *The first three columns of the projection matrix correspond to the vanishing points in* X, Y *and* Z *directions of the world frame respectively. The last column of* **P** *is the image of the world origin.*

Proof In any world coordinate system, the canonical forms of the directions along X, Y and Z axes can be written as $\mathbf{X}_w = [1,0,0,0]^T$, $\mathbf{Y}_w = [0,1,0,0]^T$ and $\mathbf{Z}_w = [0,0,1,0]^T$ respectively, and the coordinate origin is $\mathbf{O}_w = [0,0,0,1]^T$. Then from perspective projection (1.6), we have

$$\begin{cases} \mathbf{v}_x \simeq \mathbf{P}\mathbf{X}_w = \mathbf{p}_1 \\ \mathbf{v}_y \simeq \mathbf{P}\mathbf{Y}_w = \mathbf{p}_2 \\ \mathbf{v}_z \simeq \mathbf{P}\mathbf{Z}_w = \mathbf{p}_3 \\ \mathbf{v}_o \simeq \mathbf{P}\mathbf{O}_w = \mathbf{p}_4 \end{cases} \tag{1.8}$$

where '$\simeq$' denotes equality up to scale. □

1.2.2 Single View Imaging Geometry

Equation (1.6) describes the mapping of a point from 3D space to image space under perspective projection. Reversely, given an image point $\mathbf{x}$, its back-projection forms a ray $\mathbf{l} = \{\mathbf{X}|\mathbf{P}\mathbf{X} \simeq \mathbf{x}\}$ composed by the set of space points that project to $\mathbf{x}$ by the camera matrix $\mathbf{P}$. In the following text, we will present some results on the forward and back-projection of other geometric entities, such as lines, planes, and conics.

1.2.2.1 Action of Projection Matrix on Lines

A line $\mathbf{L}$ in space can be defined by two points on the line as shown in Fig. 1.2.

$$\mathbf{L} = \{\mathbf{X}(\mu) = \mathbf{X}_1 + \mu\mathbf{X}_2\} \tag{1.9}$$

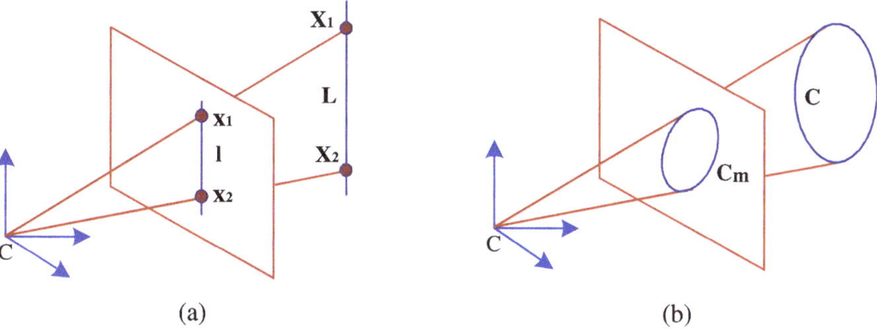

Fig. 1.2 Action of projection matrix on lines and conics. (**a**) The forward and back-projection of a line; (**b**) The forward and back-projection of a conic

where $\mathbf{X}_1$ and $\mathbf{X}_2$ are two space points on $\mathbf{L}$. The optical center and $\mathbf{L}$ define a plane, which intersect the image plane at line $\mathbf{l}$. Under the projection matrix $\mathbf{P}$, we have

$$\mathbf{l} = \{\mathbf{P}\mathbf{X}(\mu) = \mathbf{P}\mathbf{X}_1 + \mu\mathbf{P}\mathbf{X}_2\} = \{\mathbf{x}(\mu) = \mathbf{x}_1 + \mu\mathbf{x}_2\} \tag{1.10}$$

The image of $\mathbf{L}$ is a line joining the images of $\mathbf{X}_1$ and $\mathbf{X}_2$. The back-projection of the image line is a space plane passing through the optical center $\mathbf{C}$ and the line $\mathbf{l}$, which is defined by $\Pi = \mathbf{P}^T\mathbf{l}$.

1.2.2.2 Action of Projection Matrix on Planes

The projection equation (1.6) is independent of the selection of world and image frames. Suppose the xy-coordinate plane of the world frame is set on a space plane Π. Then, the coordinates of a points on the plane can be denoted as $\mathbf{X} = [x, y, 0, 1]^T$, whose image is given by

$$\lambda\mathbf{x} = \mathbf{P}\mathbf{X} = \underbrace{[\mathbf{p}_1, \mathbf{p}_2, \mathbf{p}_4]}_{\mathbf{H}} \begin{bmatrix} x \\ y \\ 1 \end{bmatrix} = \mathbf{H}\mathbf{x}_\pi \tag{1.11}$$

where $\mathbf{p}_i$ stands for the ith column of $\mathbf{P}$; $\mathbf{x}_\pi = [x, y, 1]^T$ denotes the 2D point on the space plane. Then the mapping becomes a space plane to image plane homography $\mathbf{H}$. For a point at infinity $\mathbf{X}_\infty = [x, y, z, 0]^T$, its image is called vanishing point $\mathbf{v}$, which is given by

$$\lambda\mathbf{v} = \mathbf{P}\mathbf{X}_\infty = \underbrace{[\mathbf{p}_1, \mathbf{p}_2, \mathbf{p}_3]}_{\mathbf{H}_\infty} \begin{bmatrix} x \\ y \\ z \end{bmatrix} = \mathbf{H}_\infty\bar{\mathbf{X}}_\infty \tag{1.12}$$

where $\mathbf{H}_\infty = [\mathbf{p}_1, \mathbf{p}_2, \mathbf{p}_3] = \mathbf{KR}$ is called the homography of infinite plane. $\mathbf{H}_\infty$ is of great importance in affine reconstruction.

1.2.2.3 Action of Projection Matrix on Conics

The image of a conic $\mathbf{C}$ is still a conic, as shown in Fig. 1.2. Suppose the homography from the conic plane to the image is $\mathbf{H}$, it is easy to verify that the image of the conic $\mathbf{C}_m$ is given by

$$\mathbf{C}_m = \mathbf{H}^{-T}\mathbf{C}\mathbf{H}^{-1} \tag{1.13}$$

Under the camera matrix $\mathbf{P}$, the back-projection of $\mathbf{C}_m$ is a cone defined by the optical center and the imaged conic.

$$\mathbf{Q} = \mathbf{P}^T\mathbf{C}_m\mathbf{P} \tag{1.14}$$

Specifically, the absolute conic (AC) is a conic on the ideal plane $\Pi_\infty = [0, 0, 0, 1]^T$, which can be expressed as $\Omega_\infty = \mathbf{I}_3$ or

$$[x, y, z]\Omega_\infty[x, y, z]^T = 0 \tag{1.15}$$

The image of the absolute conic (IAC) can be obtained from (1.13) by the infinite homography $\mathbf{H}_\infty$.

$$\omega = \mathbf{H}_\infty^{-T}\Omega_\infty\mathbf{H}_\infty^{-1} = \mathbf{K}^{-T}\mathbf{K}^{-1} \tag{1.16}$$

The dual image of the absolute conic (DIAC) is given by $\omega^* = \omega^{-1} = \mathbf{K}\mathbf{K}^T$. Both the IAC and DIAC are invisible imaginary conics in the image, which depend only on the camera calibration matrix $\mathbf{K}$. If the value of ω or ω^* is known, the intrinsic parameters $\mathbf{K}$ can be recovered directly by Cholesky decomposition [11]. We often refer to camera calibration as the computation of the IAC or DIAC.

1.3 Single View Metrology and Reconstruction

3D reconstruction from 2D images is a central problem of computer vision [8, 11]. The classical method for this problem is to reconstruct the metric structure of the scene from two or more images by stereo vision techniques. However, this is a difficult task due to the problem of seeking correspondences between different views. To avoid the difficulties in matching, some works were focused on reconstruction directly from a single uncalibrated image. It is well known that only one image cannot provide enough information for a complete 3D reconstruction. However, some metrical quantities can be inferred directly from a single image with a priori knowledge of geometric scene constraints. Such constraints may be expressed in terms of vanishing points or lines, co-planarity, special inter-relationship of features and camera constraints.

Many studies on single view based calibration and reconstruction are focused on structured objects or man-made environment. Caprile and Torre [3] proposed a method for camera calibration from vanishing points computed from three mutually orthogonal directions in space. Following this idea, several approaches that make use of vanishing points and lines have been proposed for either camera calibration or scene reconstruction [12, 13, 20]. Wilczkowiak *et al.* [22, 23] expanded the idea to general parallelepided structures, and use the constraints of parallelepipeds for camera calibration. Criminisi and Reid [6] studied the problem by computing 3D affine measurement from a single perspective image. Wang *et al.* [19] proposed to incorporate the symmetry property into calibration.

In this section, we will introduce some basic theory and examples on how to employ the geometric constraints for single view metrology, calibration, and 3D reconstruction.

1

1.3.1 Measurement on Space Planes

The mapping between space plane and image plane can be modeled by a homography (1.11) as

$$\lambda \mathbf{x} = \mathbf{H}\mathbf{X} \tag{1.17}$$

The homography $\mathbf{H}$ is a non-singular 3×3 homogeneous matrix with 8 degrees of freedom since it can only be defined meaningfully up to a scale factor. Each pair of space to image point correspondences in (1.17) give rise to two linear constraints on $\mathbf{H}$. Thus the homography can be uniquely determined from four point correspondences in a general position (no three points are collinear).

The homography in (1.17) can also be defined by line correspondences [16]. Suppose $\mathbf{L}$ is a line on the space plane, and $\mathbf{l}$ is the corresponding line in the image. Let $\mathbf{X}_1$ and $\mathbf{X}_2$ be two points on $\mathbf{L}$, their images are given by $\mathbf{x}_1 = s_1\mathbf{H}\mathbf{X}_1$ and $\mathbf{x}_2 = s_2\mathbf{H}\mathbf{X}_2$. Then $\mathbf{x}_1$ and $\mathbf{x}_2$ should lie on $\mathbf{l}$ and we have

$$\mathbf{l} = \mathbf{x}_1 \times \mathbf{x}_2 = (s_1\mathbf{H}\mathbf{X}_1) \times (s_2\mathbf{H}\mathbf{X}_2) = s\mathbf{H}^{-T}(\mathbf{X}_1 \times \mathbf{X}_2) = s\mathbf{H}^{-T}\mathbf{L} \tag{1.18}$$

where the scale $s = s_1 s_2 |\mathbf{H}|$; '$\times$' stands for the cross product of two vectors. In (1.18), each line correspondence can provide two constraints on the homography. Thus given four coplanar line correspondences in a general position (i.e. no three concurrent), the homography can be determined uniquely. If more than four correspondences are available, then a suitable minimization scheme can be adopted to give a more faithful estimation of the homography.

Once the homography matrix between the space and image plane is determined, an image point can be back-projected to a point on the space plane via $\mathbf{H}^{-1}$. Thus the Euclidean distance between any two points on the world plane can be simply computed from the their back-projections. This is the basic principle of plane measurements [5]. It is clear that the accuracy of this method depends greatly on the recovered homography $\mathbf{H}$. Our previous study [16] suggests that line features are more stable and can be more accurately detected than point features. The distance can also be recovered from two orthogonal sets of parallel lines via vanishing points or cross ratio [16]. In addition to the distance measurement, most other geometrical information within the space plane, such as distance from a point to a line, angle formed by two lines, area of a planar object, can also be recovered similarly from the back-projection.

Let $\mathbf{x}$, $\mathbf{l}_1$ and $\mathbf{l}_2$ be an image point and two image lines, then the corresponding point and lines in the space plane can be obtained from $\mathbf{X} = s_1\mathbf{H}^{-1}\mathbf{x}$, $\mathbf{L}_1 = s_2\mathbf{H}^T\mathbf{l}_1$ and $\mathbf{L}_2 = s_3\mathbf{H}^T\mathbf{l}_2$. Suppose the coordinates of the space point $\mathbf{X}$ and lines $\mathbf{L}_1, \mathbf{L}_2$ are $[x_s, y_s, 1]^T$, $[a_1, b_1, c_1]^T$, and $[a_2, b_2, c_2]^T$, respectively, through a simple computation, it is easy to obtain the distance from $\mathbf{X}$ to $\mathbf{L}_1$ as

$$d = \frac{|a_1 x_s + b_1 y_s + c_1|}{\sqrt{a_1^2 + b_1^2}} \tag{1.19}$$

The angle formed by $\mathbf{L}_1$ and $\mathbf{L}_2$ is given by

$$\theta = \tan^{-1}\left(\frac{a_1 b_2 - a_2 b_1}{a_1 a_2 + b_1 b_2}\right) \tag{1.20}$$

The intersection of $\mathbf{L}_1$ and $\mathbf{L}_2$ can be computed from

$$\bar{\mathbf{X}}_{12} = \left[\frac{b_1 c_2 - b_2 c_1}{a_1 b_2 - a_2 b_1}, \frac{a_2 c_1 - a_1 c_2}{a_1 b_2 - a_2 b_1}\right]^T \tag{1.21}$$

As for the area of certain planar object, we can simply extract its contour in the image, and back-project all the edge points into the world plane via the homography. Then the area of the enclosed area can be estimated by integration. For some regular objects, such as a conic or a polygon, we can fit the back-projected edge points into a conic or a polygon via least-squares or other robust estimation techniques, and then compute the area according to the corresponding formula.

1.3.2 Camera Calibration from a Single View

As we mentioned in last section, camera calibration is equivalent to finding the image of the absolute conic (IAC) or the dual image of the absolute conic (DIAC). The IAC ω is a symmetric matrix with five degrees of freedom since it is defined up to a scale. Some linear constraints on the IAC may be obtained from different camera assumptions.

Result 1.4 *For a camera with zero skew, it is easy to verify that $\omega_{12} = \omega_{21} = 0$, which can provide one linear constraint on the IAC, where ω_{ij} denotes the (i, j)th element of ω. For a camera with unit aspect ratio, we have two linear constrains on the IAC $\omega_{12} - \omega_{21} = 0$ and $\omega_{11} - \omega_{22} = 0$. For the assumption of known principal points, we may move the image center to the principal point and obtain that $\omega_{13} = \omega_{31} = 0$ and $\omega_{23} = \omega_{32} = 0$. When the camera is of zero skew with known aspect ratio κ, we have the constraints $\omega_{12} = \omega_{21} = 0$ and $\omega_{11} = \kappa^2 \omega_{22}$.*

Result 1.5 *Suppose $\mathbf{v}_1$ and $\mathbf{v}_2$ are the vanishing points of two space lines, then the included angle of the two lines is given by*

$$\theta = \arccos\left(\frac{\mathbf{v}_1^T \omega \mathbf{v}_2}{\sqrt{\mathbf{v}_1^T \omega \mathbf{v}_1}\sqrt{\mathbf{v}_2^T \omega \mathbf{v}_2}}\right)$$

If the two lines are orthogonal with each other, then we have $\mathbf{v}_1^T \omega \mathbf{v}_2 = 0$, which means that the vanishing points of the lines with orthogonal directions are conjugate with respect to the IAC.

From Result 1.5 we see that a pair of orthogonal vanishing points can provide one linear constraint on the IAC. For many structured scenarios or man-made objects, such

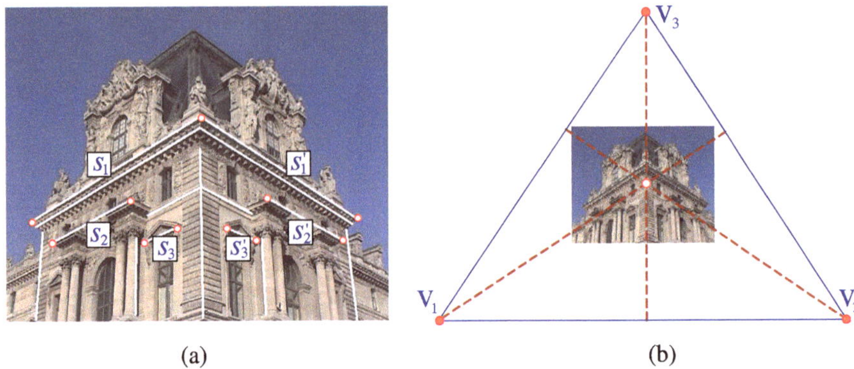

Fig. 1.3 (**a**) Three orthogonal vanishing points can be obtained from three sets of parallel lines in a single image of structured scenario; (**b**) For a camera with unit aspect ratio and zero skew, its principal point is the orthocenter of the triangle formed by the three vanishing points

as the architecture in Fig. 1.3, we can usually obtain three orthogonal vanishing points from the image of three mutually orthogonal pairs of parallel lines. Thus we have three independent constraints and the camera can be linearly calibrated under the assumption of three-parameter camera model. In the case of zero skew and unit aspect ratio, as shown in Fig. 1.3, it is proved that the principal point of the camera coincides with the orthocenter of the triangle with the three orthogonal vanishing points as vertices [11].

Due to the property of symmetry, many man-made objects usually contain some line segments with equal length or known length ratio. As shown in Fig. 1.3, several pairs of line segments have equal length, such as s_1 and s_1', s_2 and s_2', s_3 and s_3', s_4 and s_4'. It is proved in [19] that a new linear constraint on the IAC can be obtained by exploring these properties. Then it is possible to calibrate a four-parameter camera model from a single image.

When the intrinsic parameters of a camera are known, the pose of the camera can be recovered from two orthogonal vanishing points in a single view as follows. Without loss of generality, let us set the X and Y axes of the world system in line with the two orthogonal directions $\mathbf{X}_w$ and $\mathbf{Y}_w$, then from Result 1.3 we have

$$s_x\mathbf{v}_x = \mathbf{K}^{-1}\mathbf{P}\mathbf{X}_w = [\mathbf{r}_1, \mathbf{r}_2, \mathbf{r}_3, \mathbf{T}][1, 0, 0, 0]^T = \mathbf{r}_1 \tag{1.22}$$

$$s_y\mathbf{v}_y = \mathbf{K}^{-1}\mathbf{P}\mathbf{Y}_w = [\mathbf{r}_1, \mathbf{r}_2, \mathbf{r}_3, \mathbf{T}][0, 1, 0, 0]^T = \mathbf{r}_2 \tag{1.23}$$

$$s_o\mathbf{v}_o = \mathbf{K}^{-1}\mathbf{P}\mathbf{O}_w = [\mathbf{r}_1, \mathbf{r}_2, \mathbf{r}_3, \mathbf{T}][0, 0, 0, 1]^T = \mathbf{T} \tag{1.24}$$

Thus the rotation matrix can be computed from

$$\mathbf{r}_1 = \pm\frac{\mathbf{v}_x}{\|\mathbf{v}_x\|}, \qquad \mathbf{r}_2 = \pm\frac{\mathbf{v}_y}{\|\mathbf{v}_y\|}, \qquad \mathbf{r}_3 = \mathbf{r}_1 \times \mathbf{r}_2 \tag{1.25}$$

where the rotation matrix $\mathbf{R} = [\mathbf{r}_1, \mathbf{r}_2, \mathbf{r}_3]$ may have four solutions if right-handed coordinate system is adopted. While only two of them can ensure the reconstructed objects lie in front of the camera, which may be seen by the camera. In practice, if the world coordinate frame is preassigned, the rotation matrix may be uniquely determined [18]. Nevertheless, the translation vector can only be defined up to scale since we have no metric information of the given scene, which means that we can only recover the direction of the translation vector.

In practice, the orthonormal constraint should be enforced during the computation since $\mathbf{r}_1$ and $\mathbf{r}_2$ in (1.25) may not be orthogonal due to image noise. Suppose the SVD decomposition of $\mathbf{R}_{12} = [\mathbf{r}_1, \mathbf{r}_2]$ is $\mathbf{U}\Sigma\,\mathbf{V}^T$, where $\mathbf{U}$ and $\mathbf{V}$ are orthogonal matrices in the dimension of 3×3 and 2×2 respectively, Σ is a 3×2 diagonal matrix made of the two singular values of $\mathbf{R}_{12}$. Thus the best approximation to the rotation matrix can be obtained in least-square sense from

$$\mathbf{R}_{12} = \mathbf{U}\begin{bmatrix} 1 & 0 \\ 0 & 1 \\ 0 & 0 \end{bmatrix}\mathbf{V}^T \tag{1.26}$$

since a rotation matrix should have unit singular values.

1.3.3 Measurement in 3D Space

The key point in 3D measurement is the recovery of camera projection matrix $\mathbf{P}$. If both intrinsic and extrinsic parameters of a camera are known, then the projection matrix can be easily computed from $\mathbf{P} = \mathbf{K}[\mathbf{R}\,\mathbf{T}]$.

For uncalibrated cameras, the projection matrix may be recovered from scene constraints. As noted in Result 1.3, the four columns of the projection matrix correspond to the three vanishing points of the world axes and the image of the world origin. Thus the projection matrix can be written as

$$\mathbf{P} = [s_x\mathbf{v}_x, s_y\mathbf{v}_y, s_z\mathbf{v}_z, s_o\mathbf{v}_o] \tag{1.27}$$

where s_x, s_y, s_z, s_o are four unknown scalars. In the following text, we will show how to determine these unknowns.

Proposition 1.1 *Given a set of three mutually orthogonal vanishing points $\mathbf{v}_x$, $\mathbf{v}_y$ and $\mathbf{v}_z$ under a certain world system, the scalars s_x, s_y, and s_z in (1.27) may be uniquely determined if the camera is assumed to be a three-parameter-model.*

Please refer to [18] for the proof. Here we just give a simple analysis. Let

$$\mathbf{M} = [s_x\mathbf{v}_x, s_y\mathbf{v}_y, s_z\mathbf{v}_z] = [\mathbf{v}_x, \mathbf{v}_y, \mathbf{v}_z]\begin{bmatrix} s_x & & \\ & s_y & \\ & & s_z \end{bmatrix} \tag{1.28}$$

1

be the first three columns of the projection matrix, and let $\mathbf{C} = \mathbf{M}\mathbf{M}^T$, then we have

$$\mathbf{C} = \mathbf{M}\mathbf{M}^T = [\mathbf{v}_x, \mathbf{v}_y, \mathbf{v}_z]\begin{bmatrix} s_x^2 & & \\ & s_y^2 & \\ & & s_z^2 \end{bmatrix}[\mathbf{v}_x, \mathbf{v}_y, \mathbf{v}_z]^T \tag{1.29}$$

On the other hand, we know that $\mathbf{M} = \mathbf{KR}$ from (1.6). Thus

$$\mathbf{C} = \mathbf{M}\mathbf{M}^T = \mathbf{KRR}^T\mathbf{K}^T = \mathbf{KK}^T = \omega^* \tag{1.30}$$

where ω^* is the DIAC which can be obtained from the three vanishing points as shown in last subsection. Thus the three scalars can be obtained by incorporating (1.29) and (1.30).

$$\begin{bmatrix} s_x^2 & & \\ & s_y^2 & \\ & & s_z^2 \end{bmatrix} = [\tilde{\mathbf{v}}_x, \tilde{\mathbf{v}}_y, \tilde{\mathbf{v}}_z]^{-1}\omega^*[\tilde{\mathbf{v}}_x, \tilde{\mathbf{v}}_y, \tilde{\mathbf{v}}_z]^{-T} \tag{1.31}$$

However, the computed scalars are not unique, since $-s_x$, $-s_y$, and $-s_z$ also hold true for (1.31). Thus there are eight solutions that correspond to different selections of the signs in world coordinate axes. This ambiguity can be solved practically under a given world frame [18]. The last column of the projection matrix $\mathbf{p}_4$ corresponds to the translation vector of the camera to the world origin. Since there is no absolute measurement about the scene geometry, the projection matrix is only defined up to a scale, the value of s_o may be set freely. Therefore, the projection matrix is determined under the specific world system.

Proposition 1.2 *In case of a four-parameter camera model with zero skew, the camera projection matrix* $\mathbf{P}$ *can be uniquely determined from the homography* $\mathbf{H}$ *of a space plane and the vanishing point* $\mathbf{v}_z$ *in the vertical direction of the plane.*

Proof Suppose the xy-coordinate plane of world frame is set on the space plane whose homography to the image is defined by $\mathbf{H} = [\mathbf{p}_1, \mathbf{p}_2, \mathbf{p}_4]$. From (1.27) and (1.28) we can obtain the first three columns of the projection matrix as $\mathbf{M} = [\mathbf{p}_1, \mathbf{p}_2, s_z\mathbf{v}_z]$ with only one unknown scalar s_z. Then we have

$$\mathbf{C} = \mathbf{M}\mathbf{M}^T = \mathbf{p}_1\mathbf{p}_1^T + \mathbf{p}_2\mathbf{p}_2^T + s_z^2\mathbf{v}_z\mathbf{v}_z^T = \mathbf{KK}^T = \omega^* \tag{1.32}$$

Since the zero skew assumption can provide one constraint on the DIAC, from which the scalar s_z can be solved. □

Many structured objects can in general be taken as in the shape of a cuboid. A simple way is to assume the world system on the cuboid, as show in Fig. 1.4, and recover the camera projection matrix with respect to the world frame. Suppose the coordinates of a space plane is Π_i, i.e. $\Pi_i^T\mathbf{X} = 0$, with $\mathbf{X} = [x, y, z, 1]^T$. Then for an image point $\mathbf{m}_j$ on the plane, its corresponding space point $\mathbf{X}_j$ can be easily computed by the intersection of the back-projected line and the plane as:

$$\begin{cases} s_j\mathbf{x}_j = \mathbf{PX}_j \\ \Pi_i^T\mathbf{X}_j = 0 \end{cases} \tag{1.33}$$

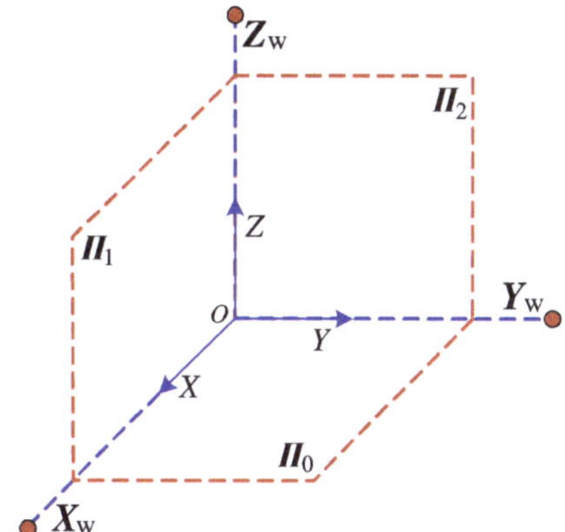

Fig. 1.4 The cuboid shape of a structured object and the selection of world coordinate system. $\mathbf{X}_w = [1, 0, 0, 0]^T$, $\mathbf{Y}_w = [0, 1, 0, 0]^T$, and $\mathbf{Z}_w = [0, 0, 1, 0]^T$ are the directions of the three axes. The coordinates of the three planes are given as $\Pi_0 = [0, 0, 1, 0]^T$, $\Pi_1 = [0, 1, 0, 0]^T$, $\Pi_2 = [1, 0, 0, 0]^T$

Similarly, for an image line $\mathbf{l}$, its back-projection is a plane $\Pi = \mathbf{P}^T \mathbf{l}$; for a conic $\mathbf{C}$ in the image, its back-projection is a view cone $\mathbf{Q} = \mathbf{P}^T \mathbf{C} \mathbf{P}$. Their corresponding space coordinates can also be computed easily from the intersection of the back-projection and the space plane.

In Fig. 1.4, the coordinates of plane Π_0, Π_1, and Π_2 are obvious, while the coordinates of most other planar surfaces can be recovered from the scene constraints with respect to the three base planes [6]. Let us take the plane Π_0 as an example.

Proposition 1.3 *The coordinates of plane Π_i parallel to Π_0 can be retrieved if a pair of corresponding points on the two planes in the vertical direction can be obtained from the image.*

Proof To recover the coordinates of Π_i is equivalent to retrieving the distance z_0 between Π_0 and Π_i. Suppose $\mathbf{X} = [x_0, y_0, 0, 1]^T$ and $\mathbf{X}' = [x_0, y_0, z_0, 1]^T$ are the pair of corresponding points on the two planes, with three unknowns x_0, y_0, z_0, as shown in Fig. 1.5. Their corresponding images are $\mathbf{x} = [u_x, v_x, 1]^T$ and $\mathbf{x}' = [u'_x, v'_x, 1]^T$ respectively. Then the three unknowns can be easily computed from

$$\begin{cases} s_1 \mathbf{x} = \mathbf{P}\mathbf{X} \\ s_2 \mathbf{x}' = \mathbf{P}\mathbf{X}' \end{cases} \tag{1.34}$$

Thus we obtain the coordinates of plane $\Pi_i = [0, 0, 1, -z_0]^T$. We can also recover the height of an object on the plane Π_0 in the same way. □

Proposition 1.4 *Suppose an arbitrary plane Π_a intersect Π_0 at line $\mathbf{L}$ as shown in Fig. 1.5, then the coordinates of plane Π_a can be determined from the images of a pair of parallel lines on the plane.*

Proof Since $\mathbf{L}$ lies on the plane $\Pi_0 = [0, 0, 1, 0]^T$, its coordinates can be easily computed. Suppose $\mathbf{L} = [a, b, 0, d]^T$, Π_v is the plane passing through $\mathbf{L}$ and orthogonal to Π_0, then

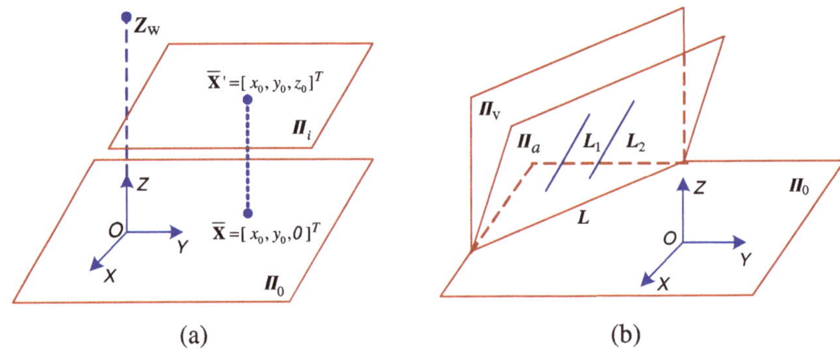

Fig. 1.5 Measurement in 3D space. (**a**) A pair of parallel planes and corresponding points between the two planes along vertical direction $\mathbf{Z}_w$; (**b**) A reference plane Π_0 and a pencil of planes passing through a line **L** on the reference plane

Π_v must have the same coordinates as **L**. All the planes passing through **L** form a pencil, and the pencil can be expressed as $\Pi_a = \Pi_v + \lambda\Pi_0 = [a, b, \lambda, d]^T$, with λ the only unknown parameter here.

Let us denote the parallel lines in space and their corresponding images as $\mathbf{L}_1$, $\mathbf{L}_2$ and $\mathbf{l}_1$, $\mathbf{l}_2$ respectively. The back projection of $\mathbf{l}_1$, $\mathbf{l}_2$ form two space planes $\Pi_{b1} = \mathbf{P}^T\mathbf{l}_1$ and $\Pi_{b2} = \mathbf{P}^T\mathbf{l}_2$. Denote the normal vector of plane Π_a, Π_{b1}, Π_{b2} as $\mathbf{n}_a$, $\mathbf{n}_{b1}$, $\mathbf{n}_{b2}$. Then the direction vector of $\mathbf{L}_1$, $\mathbf{L}_2$ can be computed from $\mathbf{n}_{L1} = \mathbf{n}_a \times \mathbf{n}_{b1}$, $\mathbf{n}_{L2} = \mathbf{n}_a \times \mathbf{n}_{b2}$. Then, λ can be easily calculated from the parallelism of the two lines.

The scalar and the plane coordinates can also be recovered in a similar way if other prior information in the arbitrary plane is available, such as two orthogonal lines, a known point, etc. □

Other geometrical entities, such as angle formed by two lines or two planes, volume and surface area of some regular and symmetric objects, etc., can also be recovered by combining the scene constraints [16, 20]. Most man-made objects, especially architectures, are usually composed of many pieces of planar surfaces. If the world coordinates of each surface can be obtained, then all the 3D information on the surface can be recovered accordingly. Thus the whole object is assembled by merging the planar patches into a 3D structure.

1.3.4 Examples of Single View Reconstruction

Many single view measurement, calibration, and reconstruction results can be found in [6, 18–20]. Here we will present two reconstruction results.

The first test is an image of a Church in Valbonne, as shown in Fig. 1.6, which is downloaded from the Visual Geometry Group of the University of Oxford. The image resolution is 512 × 768. During the test, we use Canny edge detector to detect the edge points and

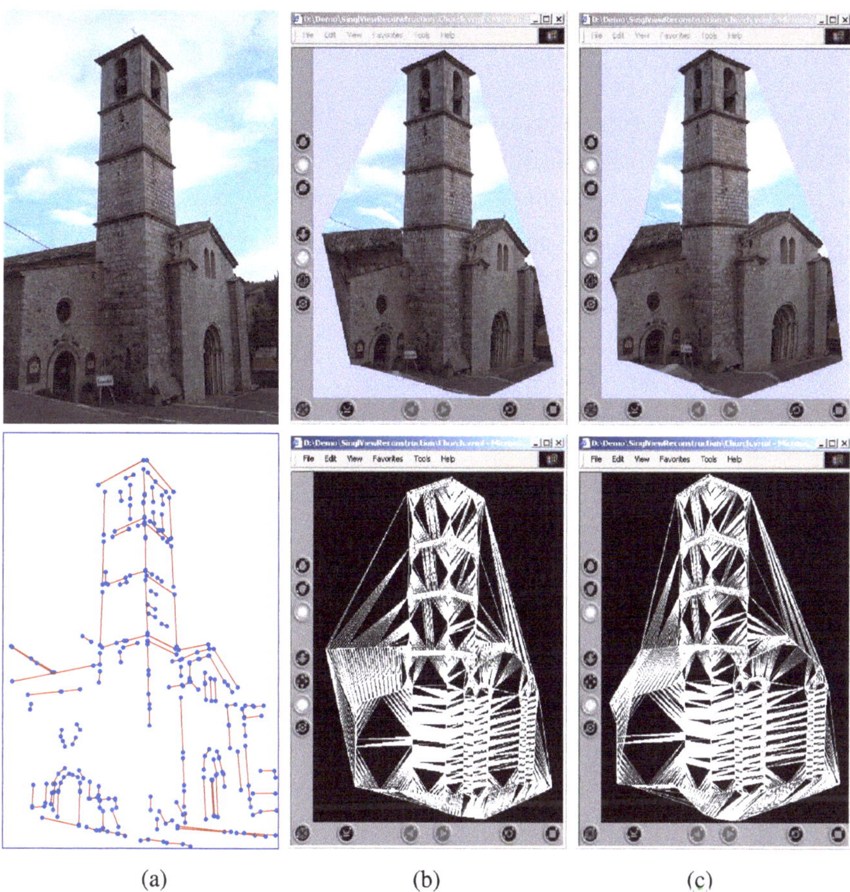

Fig. 1.6 Reconstruction result of a church in Valbonne. (**a**) One image of the church and the detected line segments; (**b**) & (**c**) The reconstructed structure of the church under different viewpoints with texture mapping and the corresponding triangulated wireframes

use Hough transform to fit the detected points into straight lines. Then the vanishing points are computed from the parallel lines using a maximum likelihood estimation [13]. We adopt the method in [19] to calibrate the camera, recover the projection matrix and the 3D structure. The reconstruction result under different viewpoint with texture mapping and the corresponding triangulated wireframes are shown in Fig. 1.6.

The second example is to recover the structure of a building site with several architectures. The site is the University Square of the Chinese University of Hong Kong, where there are four buildings around the square as shown in Fig. 1.7. All images here are taken by a Canon PowerShot G3 digital camera with the resolution of 1024×768. There are 16 images of the University Library and 12 images of the Sui-Loong Pao Building as shown in Fig. 1.7.

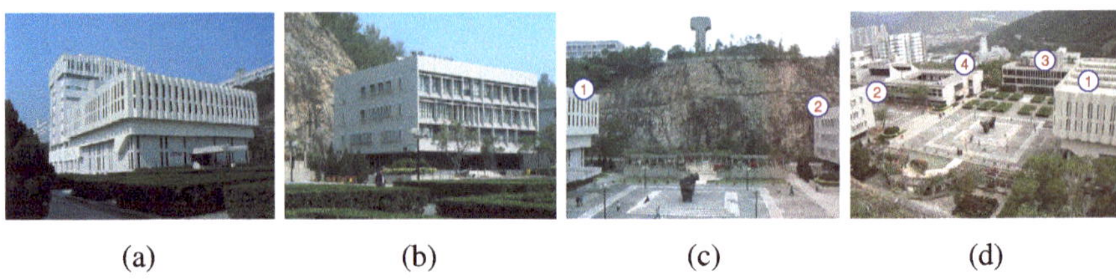

Fig. 1.7 Some image of CUHK University Square. (**a**) One sample image of the University Library sequence; (**b**) One sample image of the Sui-Loong Pao Building sequence; (**c**) & (**d**) Two birdview images of the University Square, where the building No. 1 is the University Library, No. 2 is the Sui-Loong Pao Building, No. 3 is the University Administration Building, No. 4 is the Institute of Chinese Studies

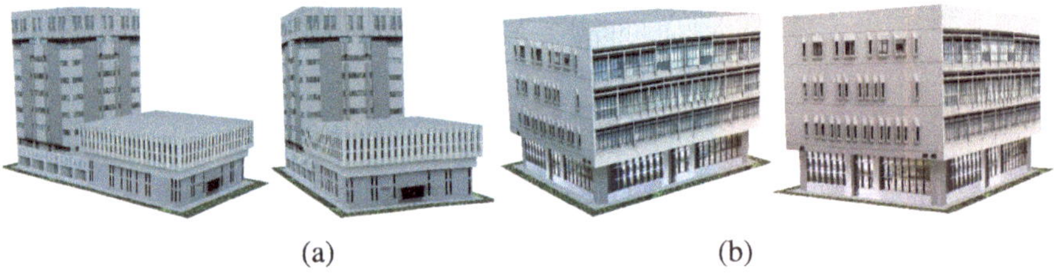

Fig. 1.8 The reconstructed models of the University Library (**a**) and the Sui-Loong Pao Building (**b**) under different viewpoints with texture mapping

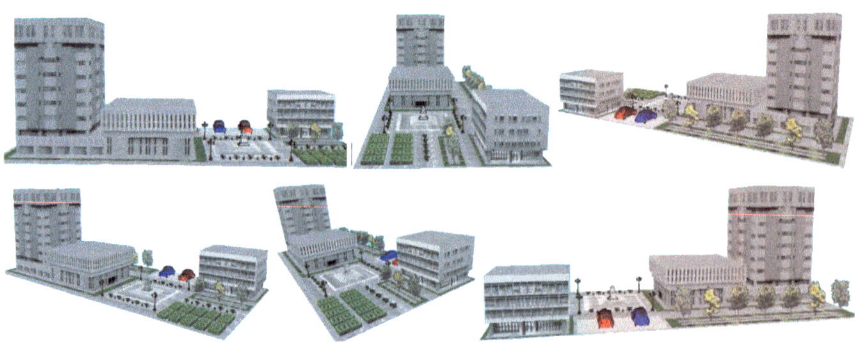

Fig. 1.9 The reconstructed virtue model of the University Square under different viewpoints

We recover the structures of the two buildings separately by the method proposed in [18]. The reconstructed models from different viewpoints are shown in Fig. 1.8, where the mapped textures to the model are slightly modified from their original images so as to remove some occlusions. We then use our study on visual metrology [20] to recover the scale, orientation and relative position of the two buildings and register them to the same coordinate frame. The final model of the University Square is shown in Fig. 1.9. Here we only reconstructed two buildings of the site. Some virtual objects, such as cars, trees, flowers, square lamp, etc. are registered and inserted into the model. Compared with the model generated by some commercial CAD softwares, the method only needs a few human interactions and the result seems more accurate and realistic since it is directly computed from the images.

1.4 Two-View Geometry and 3D Reconstruction

The application of projective geometry in computer vision is well known in structure from motion (SfM). Specifically, when only two views of the scenario are assumed, the problem is usually referred to stereo vision.

1.4.1 Epipolar Geometry and Fundamental Matrix

Epipolar geometry is the intrinsic geometry between two views, which is essentially the geometrical property of the image planes and the pencil of planes passing through the baseline of the two cameras, as shown in Fig. 1.10. The geometry is motivated by searching correspondences between the two views.

Epipolar plane: An epipolar plane is a plane passing through the two camera centers, denoted as Π in Fig. 1.10. It is a pencil of planes that have the baseline as axis, which is the line joining the two camera centers.

Epipole: An epipole is an intersection point of the baseline and image plane, denoted as $\mathbf{e}$ and $\mathbf{e}'$ for the left and right epipoles. Clearly, $\mathbf{e}$ is the projection of the right camera center in the left image, and $\mathbf{e}'$ is the image of the left optical center in the right view.

Epipolar line: An epipolar line is the intersection of an epipolar plane with the image plane, denoted as $\mathbf{l}$ and $\mathbf{l}'$ in Fig. 1.10. Obviously, the projection of any point on the epipolar plane lies on the epipolar lines. All epipolar lines intersect at the epipole.

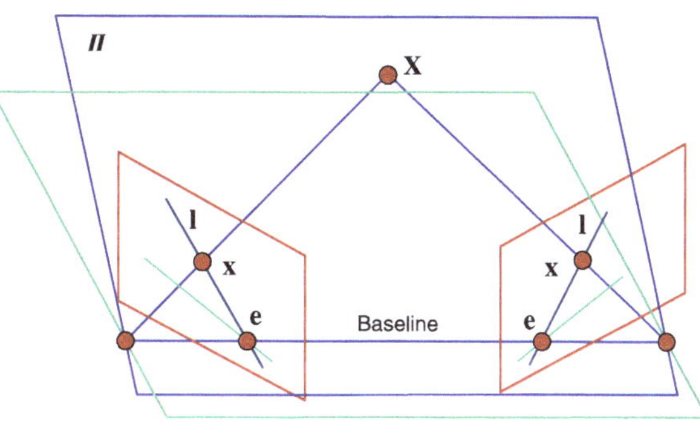

Fig. 1.10 Epipolar geometry between two views. The baseline intersect each image at the epipole; Any plane passing through the baseline is called epipolar plane; The epipolar plane intersect each image at the epipolar line. All epipolar lines pass through the epipole

1

Fundamental matrix: The fundamental matrix is the algebraic representation of epipolar geometry. For any pair of corresponding points $\mathbf{x} \leftrightarrow \mathbf{x}'$ in the two images, the fundamental matrix $\mathbf{F}$ satisfies

$$\mathbf{x}'^T \mathbf{F} \mathbf{x} = 0 \tag{1.35}$$

The fundamental matrix is a 3×3 homogeneous matrix defined up to scale. However, it has only 7 fundamental matrix since the rank of fundamental matrix $rank(\mathbf{F}) = 2$, which results in the constraint $\det(\mathbf{F}) = 0$ and removes one degree of freedom. In (1.35), each pair of points can provide one constraint on $\mathbf{F}$, thus the fundamental matrix can be solved from only 7 point correspondences for a nonlinear solution, or linearly from at least 8 pairs of correspondences. The fundamental matrix is closely related with epipolar geometry.

Result 1.6 *Suppose* $\mathbf{F}$ *is the fundamental matrix of two views* I *and* I', $\mathbf{x} \leftrightarrow \mathbf{x}'$ *is a pair of corresponding points,* $\mathbf{e}$ *and* $\mathbf{e}'$ *are epipoles in the two image. Then the two epipoles satisfy* $\mathbf{F}\mathbf{e} = 0$, $\mathbf{F}^T\mathbf{e}' = 0$. *The epipolar line corresponding to* $\mathbf{x}$ *in the other image is* $\mathbf{l}' = \mathbf{F}\mathbf{x}$, *and the epipolar line corresponding to* $\mathbf{x}'$ *is* $\mathbf{l} = \mathbf{F}^T\mathbf{x}'$.

Result 1.7 *Suppose the projection matrices of the two views are* $\mathbf{P}$ *and* $\mathbf{P}'$, *then the fundamental matrix is given by* $\mathbf{F} = [\mathbf{e}']_\times \mathbf{P}'\mathbf{P}^+$, *where* $\mathbf{P}^+$ *is the pseudo-inverse of* $\mathbf{P}$, $\mathbf{e}'$ *is the epipole which satisfies* $\mathbf{e}' = \mathbf{P}'\mathbf{C}$ *with* $\mathbf{P}\mathbf{C} = 0$. *Specifically, for two general cameras* $\mathbf{P} = \mathbf{K}[\mathbf{I}|\mathbf{0}]$, $\mathbf{P}' = \mathbf{K}'[\mathbf{R}|\mathbf{T}]$, *we have*

$$\mathbf{F} = \mathbf{K}'^{-T}[\mathbf{T}]_\times \mathbf{R}\mathbf{K}^{-1} = [\mathbf{K}'\mathbf{T}]_\times \mathbf{K}'\mathbf{R}\mathbf{K}^{-1} = \mathbf{K}'^{-T}\mathbf{R}\mathbf{K}^T[\mathbf{K}\mathbf{R}^T\mathbf{T}]_\times$$

When the cameras are calibrated and images are normalized as $\hat{\mathbf{x}} = \mathbf{K}^{-1}\mathbf{x}$, $\hat{\mathbf{x}}' = \mathbf{K}'^{-1}\mathbf{x}'$, we have a similar definition as (1.35)

$$\hat{\mathbf{x}}'^T \mathbf{E} \hat{\mathbf{x}} = 0 \tag{1.36}$$

The matrix $\mathbf{E}$ is named as essential matrix. Clearly, the essential matrix and fundamental matrix are related by $\mathbf{E} = \mathbf{K}'^T\mathbf{F}\mathbf{K}$. Under normalized images, the two cameras are given by $\mathbf{P} = [\mathbf{I}|\mathbf{0}]$, $\mathbf{P}' = [\mathbf{R}|\mathbf{T}]$. Then from Result 1.7, we can immediately have

$$\mathbf{E} = [\mathbf{T}]_\times \mathbf{R} = \mathbf{R}[\mathbf{R}^T\mathbf{T}]_\times \tag{1.37}$$

The essential matrix is a homogeneous matrix defined up to an overall scale ambiguity. Thus $\mathbf{E}$ has only five degrees of freedom, since both $\mathbf{R}$ and $\mathbf{T}$ have three degrees of freedom.

1.4.2 Three Dimensional Reconstruction

3D reconstruction is one of the main goals in computer vision. Suppose we have sufficient image correspondences $\mathbf{x}_i \leftrightarrow \mathbf{x}'_i$, we want to recover the corresponding 3D points $\mathbf{X}_i$ in space and the camera matrices.

As discussed in last subsection, a fundamental matrix can be recovered from point correspondences. Given the fundamental matrix, a pair of camera projection matrices can be determined up to a projective transformation in $\mathbb{P}^3$. Specifically, a pair of canonical cameras can be chosen as $\mathbf{P} = [\mathbf{I}|\mathbf{0}]$ and $\mathbf{P}' = [[\mathbf{e}']_\times \mathbf{F}|\mathbf{e}']$, or more generally, $\mathbf{P}' = [[\mathbf{e}']_\times \mathbf{F} + \mathbf{e}'\mathbf{v}^T|\lambda\mathbf{e}']$, where $\mathbf{v}$ is an arbitrary 3-vector, and λ is a nonzero scale. See [11, 14] for proof.

After recovering the projection matrices, the 3D structure of any image correspondences can be computed via triangulation.

$$\begin{cases} \lambda_i \mathbf{x}_i = \mathbf{P}\mathbf{X}_i \\ \lambda'_i \mathbf{x}'_i = \mathbf{P}'\mathbf{X}_i \end{cases} \tag{1.38}$$

As shown in Fig. 1.10, the space point $\mathbf{X}_i$ is the intersection of two rays back-projected from the two image points $\mathbf{x}_i$ and $\mathbf{x}'_i$. This is a projective reconstruction in $\mathbb{P}^3$ since nothing is known of the calibration of the cameras nor the relative motion of the two cameras. The selection of camera projection matrices is not unique, thus the recovered structure $\{\mathbf{X}_i\}$ is different. Any two reconstructions are defined up to a nonsingular transformation matrix.

Suppose $(\mathbf{P}, \mathbf{P}', \{\mathbf{X}_i\})$ is one computed reconstruction, and $(\mathbf{P}_r, \mathbf{P}'_r, \{\mathbf{X}_{ri}\})$ is a real Euclidean reconstruction, i.e. $\{\mathbf{X}_{ri}\}$ is the true structure in Euclidean space, which is imaged as $\{\mathbf{x}_i \leftrightarrow \mathbf{x}'_i\}$ by two real cameras $\mathbf{P}_r$ and $\mathbf{P}'_r$. Then there exists a nonsingular matrix $\mathbf{H}$ that satisfies

$$\mathbf{P}_r = \mathbf{P}\mathbf{H}^{-1}, \qquad \mathbf{P}'_r = \mathbf{P}'\mathbf{H}^{-1}, \qquad \mathbf{X}_{ri} = \mathbf{H}\mathbf{X}_i \tag{1.39}$$

If $\mathbf{H}$ is a projective transformation matrix, we call $(\mathbf{P}, \mathbf{P}', \{\mathbf{X}_i\})$ a projective reconstruction; If $\mathbf{H}$ is an affine transformation matrix, then the reconstruction is in an affine coordinate frame, we call it affine reconstruction. If $\mathbf{H}$ is a similarity transformation matrix, we call $(\mathbf{P}, \mathbf{P}', \{\mathbf{X}_i\})$ a similarity reconstruction or metric reconstruction. In most literatures, it is also referred as Euclidean reconstruction, since the Euclidean properties are preserved in similarity space.

There are usually two ways to obtain the Euclidean reconstruction: one is a stratified reconstruction, the other is a direct metric reconstruction. The stratified approach starts from a projective reconstruction which can be recovered directly from the fundamental matrix, then the solution is refined to an affine and finally a metric reconstruction via some further information about the scene and cameras. The essence of stratification to affine reconstruction is to locate the plane at infinity by some scene constraints. The key to metric reconstruction is the identification of the absolute conic [9, 11].

When the cameras are calibrated, we can recover the metric projection matrices directly from the essential matrix, and a metric reconstruction can be obtained in a straightforward manner. As shown in [11], there are four possible solutions in this case, while only one solution (the true reconstruction) guarantees that the reconstructed points lie in front of both cameras. The false reconstructions can be ruled out by a sample point.

1.5 Reconstruction of Structured Scenes from Two Images

A lot of efforts on two view reconstruction have been spent on the development of high accuracy and complete modelling of complex scenes based on matching primitive features. However, many methods do not observe the geometric context and constraints arising from the scenes. In this section, we focus on the reconstruction of structured scenes from uncalibrated images and present a practical reconstruction method by incorporating the information of points, lines, planes, and the geometrical constraints [17].

There are many related studies in the literature that utilize the scene constraints for architecture reconstruction [2, 23]. Werner and Zisserman [21] proposed to generate the planar model of the principal scene planes by plane sweeping. Baillard and Zisserman [1] utilized inter-image homographies to validate and estimate the plane models. Bartoli and Sturm [2] proposed to describe the geometric relationship between points and planes by multi-coplanarity constraints, and a maximum likelihood estimator that incorporates the constraints and structures was used in a bundle adjustment manner. Some commercial systems, such as Facade [7] and Photobuilder [4] can produce very realistic results. However, those systems usually require a lot of human interactions.

In this section, we will present a practical hybrid method for reconstruction of structured scenes from two uncalibrated images. The method is based on an initial estimation of principal homographies from 2D point matches. Then match line segments between images and refine the homography.

1.5.1 Plane Detection Strategy

The strategy is based on an initial point matching result. We first estimate the principal planar homography by a recursive RANSAC method. Then find the line correspondences guided by the estimated homography and refine the homography by incorporating both point and line features on the plane.

1.5.1.1 Coarse Estimation of Plane Homography

Seeking correspondances between images is a difficult task. Most of the available methods are found to be error prone for man-made structure scenes due to the ambiguities caused either by large homogeneous regions of texture or repeated patterns in images. Figure 1.11 shows two images of the Wadham College of Oxford. We establish 1128 initial matches using the method of correlation and relaxation [24] with about 13% of outliers. For structured scenario, most of the matches usually lie on several principal planar surfaces, and the matches on each surface are related with a planar homography. We adopt a RANSAC mechanism [10] to estimate the planar homography recursively.

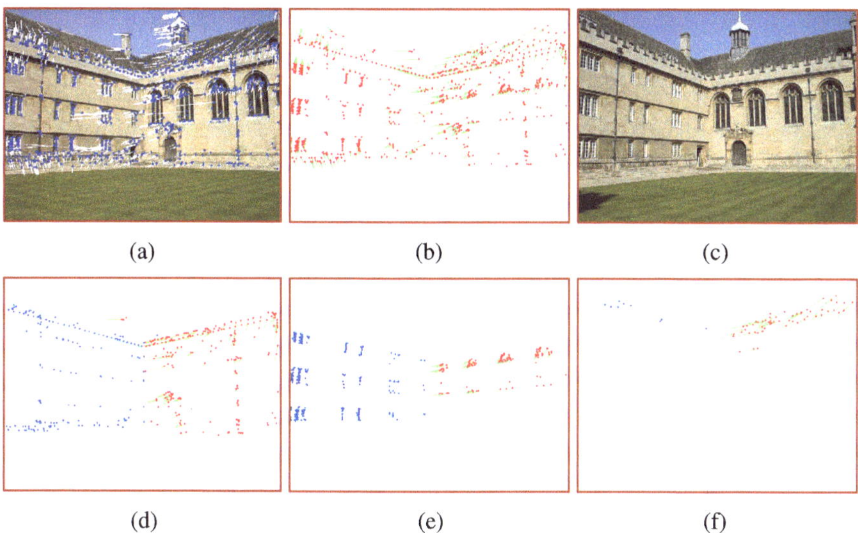

Fig. 1.11 (**a**) & (**c**) Two images of the Wadham College of Oxford with a total of 1128 initial matches shown in the first view with disparities to the second view; (**b**) The detected correct matches on principal planes shown in the first view with disparities; (**d**) The supporting matches of the left and right walls; (**e**) The supporting matches of the left and right windows; (**f**) The supporting matches of the left and right roofs

During iteration, planes are hypothesized from a randomly selected sample of four pairs of point matches (if a reasonable fundamental matrix can be computed from the initial matches, then three pairs of matches are enough to generate the planar homography [18]). The plane which is best supported by those point pairs which have not been assigned to previous planes is selected. It should be noted that this method can only detect the principal surfaces of the object with relatively large number of initial matches. Figure 1.11 shows a result of the plane detection based on the initial matches. The supporting matches for each plane are shown in different colors with their positions in the first view, while the outliers and the features not associated with the detected planes are eliminated. Totally 5 planes (which correspond to the left and right wall, the right roof, and the windows of left and right walls) are detected automatically.

1.5.1.2
Homography-Guided Line Matching

Line matching is a difficult problem since there are no strong disambiguating geometric constraints available and the topological connections between line segments are often lost during segmentation. Since most of the segments of a structured object lie on the detected principal surfaces, the matching between these line segments can be obtained using the homographies.

1

We apply the Canny edge detector and orthogonal regression algorithm to fit the line segments in images [15]. Suppose the total number of extracted line segments in image 1 and image 2 are n_1 and n_2 respectively. For a segment $\mathbf{l}'_j$ $(j = 1, \dots, n_2)$ in the second view, we can map it into the first view as $\hat{\mathbf{l}}_j \simeq \mathbf{H}_k^T \mathbf{l}'_j$, where, $\mathbf{H}_k$ is the homography induced by the kth plane of the object. Then compare some relations between $\hat{\mathbf{l}}_j$ and $\mathbf{l}_i$ $(i = 1, \dots, n_1)$ according to the following criteria

$$\begin{cases} angle(\mathbf{l}_i, \hat{\mathbf{l}}_j) < \varepsilon_1 \\ dist^2(\mathbf{l}_i, \hat{\mathbf{l}}_j) < \varepsilon_2 \\ overlap(\mathbf{l}_i, \hat{\mathbf{l}}_j) > \varepsilon_3 \end{cases} \tag{1.40}$$

and adopt a winner-take-all strategy to select the matching candidates that correspond to the plane. Here, $angle(\bullet, \bullet)$ denotes the smaller angle between two lines, $dist^2(\bullet, \bullet)$ denotes the sum of square of the Euclidean distance between the two endpoints of the first line segment to the second line, $overlap(\bullet, \bullet)$ denotes the length of overlap of the two line segments.

Most of the matching candidates obtained above are correct (i.e. lie on the images of the kth plane in space). However, a few of them may lie outside the plane (referred as incorrect matches here). For two pairs of matched lines $\mathbf{l}_i \leftrightarrow \mathbf{l}'_i$ and $\mathbf{l}_j \leftrightarrow \mathbf{l}'_j$, if they correspond to two coplanar lines in space, as shown in Fig. 1.12, then their intersections $\mathbf{x}_{ij}$ and $\mathbf{x}'_{ij}$ must satisfy

$$\mathbf{x}'_{ij} \simeq \mathbf{H}_k \mathbf{x}_{ij} \tag{1.41}$$

For each pair of obtained candidates, we compute its intersection points with the remaining of line pairs. If more than 50% of the intersections satisfy (1.41), then this pair is considered to be correct. Otherwise, eliminate this match from the list. In the same way, we can obtain all correct line matches on each plane. Figure 1.13 shows the matching results corresponding to every detected space plane.

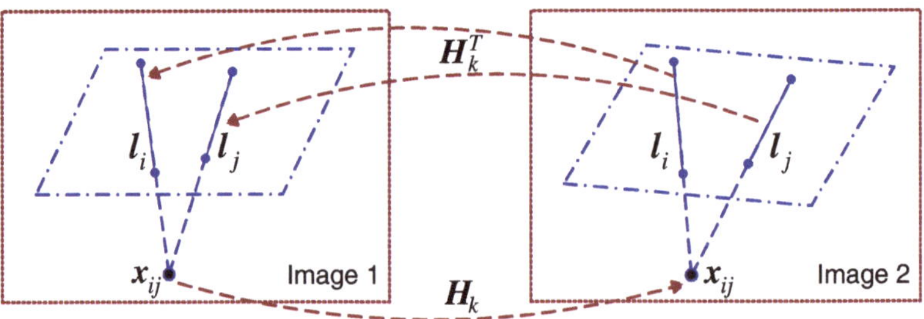

Fig. 1.12 The intersection of two coplanar lines still lies on the plane, and its images in two views satisfy the homography constraint

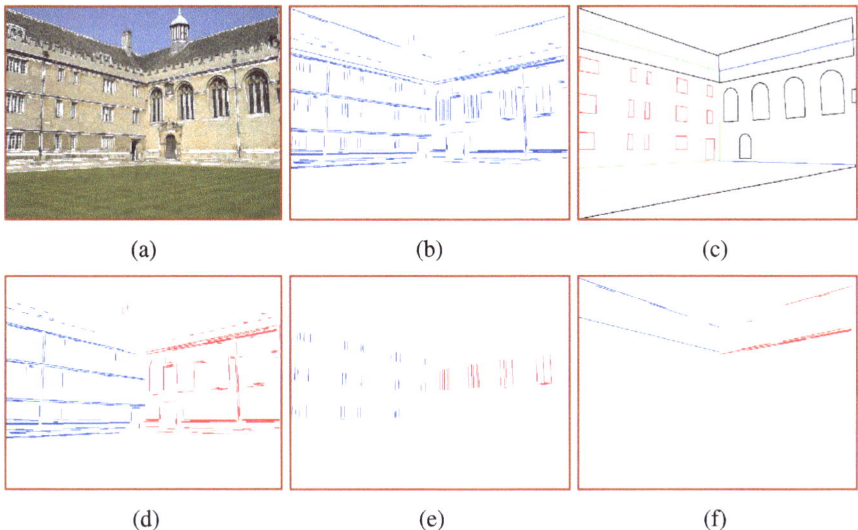

Fig. 1.13 (**a**) The initial detected 328 line segments in the first view; (**b**) All correct line matches on the principal planes; (**c**) The detected contours of the 6 principal planes; (**d**) The correct line matches detected on the left and right walls; (**e**) The correct line matches detected on the left and right windows; (**f**) The correct line matches detected on the left and right roofs

1.5.1.3 Homography Refinement

After obtaining the correct matches of feature points and line segments on the detected planes, the homography can be re-estimated by incorporating the information of both points and lines, since line features are more stable and can be more precisely detected than point features.

$$\begin{cases} \mathbf{x}'_i \simeq \mathbf{H}_k \mathbf{x}_i & (i = 1, \ldots, m) \\ \mathbf{l}_j \simeq \mathbf{H}_k^T \mathbf{l}'_j & (j = 1, \ldots, n) \end{cases} \tag{1.42}$$

where m and n are the numbers of matched points and lines on the kth plane in space. The above equation system can provide $2(m+n)$ linear constraints on the 8 entities of homography. Thus $\mathbf{H}_k$ can be computed by least-squares estimation. The solution may be further optimized by minimizing the following cost function.

$$\begin{aligned} J(\mathbf{H}_k) = {} & \frac{1}{m} \sum_{i=1}^{m} \left(d^2(\mathbf{x}'_i, \mathbf{H}_k \mathbf{x}_i) + d^2(\mathbf{x}_i, \mathbf{H}_k^{-1} \mathbf{x}'_i) \right) \\ & + \frac{1}{2n} \sum_{j=1}^{n} \left(dist^2(\mathbf{l}_j, \mathbf{H}_k^T \mathbf{l}'_j) + dist^2(\mathbf{l}'_j, \mathbf{H}_k^{-T} \mathbf{l}_j) \right) \end{aligned} \tag{1.43}$$

where the first term is the normalized geometric re-projection error of the matched points in bilateral directions, the second term is the normalized error of the line matches in the two views, and $d^2(\bullet, \bullet)$ denotes the sum of square of the Euclidean distance between two points. Using the refined homographies, we may find more supporting point and line matches on the detected principal surfaces and refine the homography again.

After retrieving all homographies of the planar surfaces, we can compute the line of intersection between each pair of planes. Suppose $\mathbf{H}_1$ and $\mathbf{H}_2$ are homographies induced by two space planes. Let $\mathbf{l} \leftrightarrow \mathbf{l}'$ be the projections of the intersection of the two planes. Then from

$$\mathbf{l} \simeq \mathbf{H}_1^T \mathbf{l}', \qquad \mathbf{l} \simeq \mathbf{H}_2^T \mathbf{l}' \tag{1.44}$$

we have

$$(\mathbf{H}_2^{-1}\mathbf{H}_1)^T \mathbf{l} \simeq \mathbf{l} \tag{1.45}$$

Thus the intersection in the first view may be determined from the eigenvector corresponding to the real eigenvalue of matrix $(\mathbf{H}_2^{-1}\mathbf{H}_1)^T$. Actually, the matrix $\mathbf{H}_2^{-1}\mathbf{H}_1$ is the mapping from the first image onto itself. It is a planar homology which has a fixed point (vertex) and a line of fixed points (axis) [11]. The transformation has two equal and one distinct eigenvalues. The axis, which is the intersection line of the two planes, is the join of the eigenvectors corresponding to the degenerate eigenvalues. The third eigenvector corresponds to the vertex, which is the epipole in the second view.

According to the intersections of any two adjacent planes and the line matches on each plane, it is easy to obtain the contour of each detected plane. Figure 1.13 gives an example. Homography provides a one-to-one mapping of points and lines between two views. For any points or lines on the detected planes in one image, we can immediately obtain their correspondences in the other view, even if these correspondences may be occluded or lie outside the image.

1.5.2 Camera Calibration and Reconstruction

The image of the absolute conic (IAC) $\boldsymbol{\omega} = (\mathbf{K}\mathbf{K}^T)^{-1}$ depends only on the camera calibration matrix $\mathbf{K}$. It is a symmetric matrix with five degrees of freedom defined up to scale. Once $\boldsymbol{\omega}$ is computed, the intrinsic parameters of the camera can be recovered by Cholesky decomposition.

As we mentioned in last section, we can usually obtain three mutually orthogonal vanishing points for structured objects. Thus we may have the following three constraints on the IAC.

$$\begin{cases} \mathbf{v}_x^T \omega \mathbf{v}_y = 0 \\ \mathbf{v}_y^T \omega \mathbf{v}_z = 0 \\ \mathbf{v}_z^T \omega \mathbf{v}_x = 0 \end{cases} \tag{1.46}$$

Similarly, we can have the constraints on ω' in the second view. The three pairs of vanishing points $\mathbf{v}_x \leftrightarrow \mathbf{v}'_x$, $\mathbf{v}_y \leftrightarrow \mathbf{v}'_y$, $\mathbf{v}_z \leftrightarrow \mathbf{v}'_z$ lie on the plane at infinity, thus we can retrieve the infinite homography $\mathbf{H}_\infty$ from the three pairs of vanishing points together with the fundamental matrix [11]. The absolute conic also lies on the infinite plane whose images in the two views are related by $\mathbf{H}_\infty$.

$$\omega' = \mathbf{H}_\infty^{-T} \omega \mathbf{H}_\infty^{-1} \tag{1.47}$$

This means that the constraints on the IAC can be easily transferred from one view to the other via the infinite homography. It can be verified that (1.47) can provide 5 independent constraints on the IAC when the cameras have varying parameters. Therefore, the cameras can be calibrated from (1.46) and (1.47) under the assumption of zero-skew [17].

After retrieving the camera parameters, a standard structure from motion algorithm is employed to reconstruct the whole object in the Euclidean space according to the contours shown in Fig. 1.13. The reconstruction result for the Wadham College is shown in Fig. 1.14.

Fig. 1.14 Reconstruction result of the Wadham college shown from different viewpoints with texture mapping

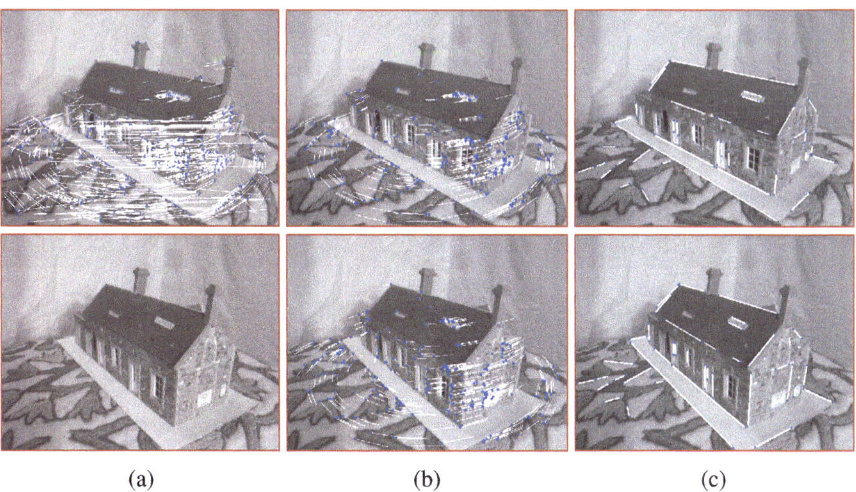

Fig. 1.15 (**a**) Two images of a model house with initial matches shown in the first view; (**b**) The supporting matches to the four principal planes (ground, roof, front and side walls); (**c**) The matching results of line segments on the four principal planes

  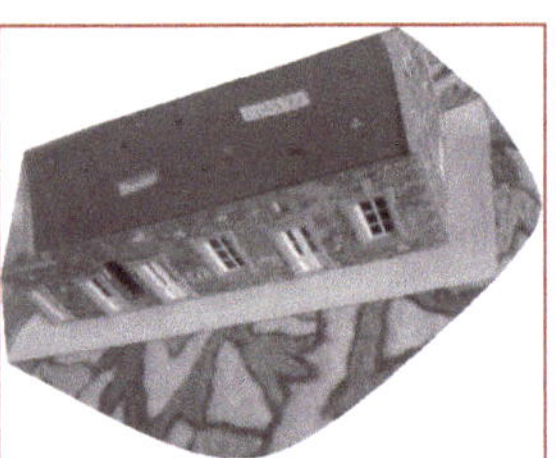

Fig. 1.16 Reconstruction result of the model house shown from different viewpoints with texture mapping

There are three mutually orthogonal planes in the scene: the ground plane, the left and the right walls. The reconstructed angle between these planes are 90.36°, 90.28°, and 89.85° respectively.

We carried another test on two images of a model house with a resolution of 768×576. As shown in Fig. 1.15, there are 549 initial matches with about 22% outliers. Using the proposed methods, four principal planes, which correspond to the ground, the roof, the front and the side walls, are detected automatically. A total of 333 pairs of supporting points and 38 pairs of matched line segments on these planes are detected as shown in Fig. 1.15. We interactively draw the contours of the house from the line matches and intersections of planes. The reconstruction result under different viewpoint with texture mapping is shown in Fig. 1.16. The three reconstructed angles between the ground and the two walls are 90.27°, 90.41°, and 89.78° respectively. We can see from all the reconstructions that they are largely consistent with the real cases, and seem very realistic.

1.6 Closure Remarks

1.6.1 Conclusion

In this chapter, we reviewed some preliminary knowledge on camera geometry in one view and two views. We also presented some application examples on calibration, visual metrology, and 3D reconstruction. There are still many problems not covered here, such as three-view geometry, the numerical computation methods of homography, camera matrix, and fundamental matrix. Interested readers may refer to [8, 11] for more details.

1.6.2 Review Questions

1. *Imaging geometry*. Give the equation of perspective projection. Describe the geometric interpretation of each column of the projection matrix. Show the form of a general camera calibration matrix **K** and meaning of each parameters.

2. *Homography.* Derive the plane to plane homography from point features and line features. Show how measurements on a plane are taken from a single view.
3. *Single view geometry.* Show the forward and backward projection equations of lines and conics. Show how a camera can be calibrated from three mutually orthogonal vanishing points.
4. *Two-view geometry.* Obtain the relationship between fundamental matrix and epipolar geometry. Show how to determine a pair of camera matrices from the fundamental matrix.
5. *Plane intersection.* Given the plane induced homographies of two space planes, and compute the intersection line of the two planes.

References

1. Baillard, C., Zisserman, A.: Automatic reconstruction of piecewise planar models from multiple views. In: Proceedings of the IEEE Conference on Computer Vision and Pattern Recognition, vol. 2, pp. 559–565 (1999)
2. Bartoli, A., Sturm, P.: Constrained structure and motion from multiple uncalibrated views of a piecewise planar scene. Int. J. Comput. Vis. **52**(1), 45–64 (2003)
3. Caprile, B., Torre, V.: Using vanishing points for camera calibration. Int. J. Comput. Vis. **4**(2), 127–140 (1990)
4. Cipolla, R., Robertson, D., Boyer, E.: Photobuilder—3D models of architectural scenes from uncalibrated images. In: Proceedings of the IEEE International Conference on Multimedia Computing and Systems, pp. 25–31. IEEE Computer Society, Washington (1999)
5. Criminisi, A., Reid, I., Zisserman, A.: A plane measuring device. Image Vis. Comput. **17**(8), 625–634 (1999)
6. Criminisi, A., Reid, I.D., Zisserman, A.: Single view metrology. Int. J. Comput. Vis. **40**(2), 123–148 (2000)
7. Debevec, P.E., Taylor, C.J., Malik, J.: Modeling and rendering architecture from photographs: A hybrid geometry- and image-based approach. In: SIGGRAPH, pp. 11–20 (1996)
8. Faugeras, O.: Three-Dimensional Computer Vision: A Geometric Viewpoint. MIT Press, Cambridge (1993)
9. Faugeras, O.: Stratification of three-dimensional vision: projective, affine, and metric representations. J. Opt. Soc. Am. A **12**, 465–484 (1995)
10. Fischler, M.A., Bolles, R.C.: Random sample consensus: A paradigm for model fitting with applications to image analysis and automated cartography. Commun. ACM **24**(6), 381–395 (1981)
11. Hartley, R.I., Zisserman, A.: Multiple View Geometry in Computer Vision, 2nd edn. Cambridge University Press, Cambridge (2004). ISBN: 0521540518
12. Heuvel, F.A.v.: 3D reconstruction from a single image using geometric constraints. ISPRS J. Photogram. Remote Sens. **53**(6), 354–368 (1998)
13. Liebowitz, D., Criminisi, A., Zisserman, A.: Creating architectural models from images. In: Proceedings of Eurographics, pp. 39–50 (1999)
14. Luong, Q.T., Viéville, T.: Canonical representations for the geometries of multiple projective views. Comput. Vis. Image Underst. **64**(2), 193–229 (1996)
15. Schmid, C., Zisserman, A.: Automatic line matching across views. In: Proceedings of the IEEE Conference on Computer Vision and Pattern Recognition, pp. 666–671 (1997)
16. Wang, G., Hu, Z., Wu, F.: Single view based measurement on space planes. J. Comput. Sci. Technol. **19**(3), 374–382 (2004)

17. Wang, G., Tsui, H.T., Hu, Z.: Reconstruction of structured scenes from two uncalibrated images. Pattern Recogn. Lett. **26**(2), 207–220 (2005)
18. Wang, G., Tsui, H.T., Wu, J.: What can we learn about the scene structure from three orthogonal vanishing points in images. Pattern Recogn. Lett. **30**(3), 192–202 (2009)
19. Wang, G., Tsui, H.T., Hu, Z., Wu, F.: Camera calibration and 3D reconstruction from a single view based on scene constraints. Image Vis. Comput. **23**(3), 311–323 (2005)
20. Wang, G., Hu, Z., Wu, F., Tsui, H.T.: Single view metrology from scene constraints. Image Vis. Comput. **23**(9), 831–840 (2005)
21. Werner, T., Zisserman, A.: New techniques for automated architecture reconstruction from photographs. In: Proceedings of the 7th European Conference on Computer Vision, Copenhagen, Denmark, vol. 2, pp. 541–555 (2002)
22. Wilczkowiak, M., Boyer, E., Sturm, P.: Camera calibration and 3D reconstruction from single images using parallelepipeds. In: Proceedings of International Conference on Computer Vision, vol. 1 (2001)
23. Wilczkowiak, M., Boyer, E., Sturm, P.: 3D modeling using geometric constraints: A parallelepiped based approach. In: Proceedings of the 7th European Conference on Computer Vision, Copenhagen, Denmark, vol. 4, pp. 221–237 (2002)
24. Zhang, Z., Deriche, R., Faugeras, O., Luong, Q.T.: A robust technique for matching two uncalibrated images through the recovery of the unknown epipolar geometry. Artif. Intell. **78**(1–2), 87–119 (1995)

Simplified Camera Projection Models 2

Abstract The chapter focuses on the approximation of full perspective projection model. We first present a review on affine camera model, including orthographic projection, weak-perspective projection, and paraperspective projection. Then, under the assumption that the camera is far away from the object with small lateral rotations, we prove that the imaging process can be modeled by quasi-perspective projection. The model is proved to be more accurate than affine model using both geometrical error analysis and experimental studies.

Everything should be made as simple as possible, but not simpler.

Albert Einstein (1879–1955)

2.1 Introduction

The modeling of imaging process is an important issue for many computer vision applications, such as structure from motion, object recognition, pose estimation, etc. Geometrically, a camera maps data from 3D space to a 2D image space. The general camera model used in computer vision is modeled by perspective projection. This is an ideal and accurate model for a wide range of existing cameras. However, the resulting equations from perspective projection are complicated and often nonlinear due to the unknown scaling factor [7]. To simplify computations, researchers have proposed many approximations to the full perspective projection.

The most common approximation includes weak-perspective projection, orthographic projection, and paraperspective projection [1]. These approximations are generalized as affine camera model [5, 9]. Faugeras [2] introduced the properties of projective cameras. Hartley and Zisserman [3] presented a comprehensive survey and in-depth analysis on different camera models. Affine camera is a zero-order (for weak-perspective) or a first-order (for paraperspective) approximation of full perspective projection. It is valid when the depth variation of the object is small compared to the distance from camera to the object. Kanatani *et al.* [4] analyzed a general form of symmetric affine camera model to mimic

G. Wang, Q.M.J. Wu, *Guide to Three Dimensional Structure and Motion Factorization*, Advances in Pattern Recognition,
DOI 10.1007/978-0-85729-046-5_2,

perspective projection and provided the minimal requirements for orthographic, weak perspective, and para-perspective simplification. The model contains two free variables that can be determined through self-calibration.

Affine assumption is widely adopted for the study of structure from motion due to its simplicity. In this chapter, we try to make a trade-off between simplicity of the affine model and accuracy of the full perspective projection model. Assuming that the camera is far away from the object with small lateral rotations, which is similar to affine assumption and is easily satisfied in practice. We propose a quasi-perspective projection model and present an error analysis of different projection models [10]. The model is proved to be more accurate than affine approximation. In the subsequent chapters of this book, we will provide some two-view properties of the model [11] and its application to structure and motion factorization [12].

The remaining part of the chapter is organized as follows. The affine projection model is reviewed in Sect. 2.2. The proposed quasi-perspective model and error analysis are elaborated in Sect. 2.3. Some experimental evaluations on synthetic data are given in Sect. 2.4.

2.2 Affine Projection Model

Under perspective projection, a 3D point $\mathbf{X}_j$ is projected onto an image point $\mathbf{x}_{ij}$ in frame i according to equation

$$\lambda_{ij}\mathbf{x}_{ij} = \mathbf{P}_i\mathbf{X}_j = \mathbf{K}_i[\mathbf{R}_i, \mathbf{T}_i]\mathbf{X}_j \tag{2.1}$$

where λ_{ij} is a non-zero scale factor, commonly denoted as the projective depth; the image point $\mathbf{x}_{ij}$ and space point $\mathbf{X}_j$ are expressed in homogeneous form; $\mathbf{P}_i$ is the projection matrix of the i-th frame; $\mathbf{R}_i$ and $\mathbf{T}_i$ are the corresponding rotation matrix and translation vector of the camera with respect to the world system; $\mathbf{K}_i$ is the camera calibration matrix of the form

$$\mathbf{K}_i = \begin{bmatrix} f_i & \varsigma_i & u_{0i} \\ 0 & \kappa_i f_i & v_{0i} \\ 0 & 0 & 1 \end{bmatrix} \tag{2.2}$$

For some precise industrial CCD cameras, we assume zero skew $\varsigma_i = 0$, known principal point $u_{0i} = v_{0i} = 0$, and unit aspect ratio $\kappa_i = 1$. Then the camera is simplified to have only one intrinsic parameter f_i.

When the distance of an object from a camera is much greater than the depth variation of the object, we may assume affine camera model. Under affine assumption, the last row of the projection matrix is of the form $\mathbf{P}_{3i}^T \simeq [0, 0, 0, 1]$, where '$\simeq$' denotes equality up to scale. Thus a general affine projection matrix for the ith view can be written as

$$\mathbf{P}_{Ai} = \begin{bmatrix} p_{11} & p_{12} & p_{13} & p_{14} \\ p_{21} & p_{22} & p_{23} & p_{24} \\ 0 & 0 & 0 & 1 \end{bmatrix} = \begin{bmatrix} \mathbf{A}_i & \bar{\mathbf{T}}_i \\ \mathbf{0}^T & 1 \end{bmatrix} \tag{2.3}$$

where, $\mathbf{A}_i \in \mathbb{R}^{2\times 3}$ is composed by the upper-left 2×3 submatrix of $\mathbf{P}_i$, $\bar{\mathbf{T}}_i$ is a translation vector. Then, the projection process (2.1) can be simplified by removing the scale factor λ_{ij}.

$$\bar{\mathbf{x}}_{ij} = \mathbf{A}_i \bar{\mathbf{X}}_j + \bar{\mathbf{T}}_i \tag{2.4}$$

Under affine projection, the mapping from space to the image is linear. One attractive attribute of affine camera model is that the mapping is independent of the translation term if relative coordinates are employed in both space and image coordinate frames.

Suppose $\bar{\mathbf{X}}_r$ is a reference point in space and $\bar{\mathbf{x}}_{ir}$ is its image in the ith frame. Then, we have $\bar{\mathbf{x}}_{ir} = \mathbf{A}_i \bar{\mathbf{X}}_r + \bar{\mathbf{T}}_i$. Let us denote

$$\bar{\mathbf{x}}'_{ij} = \bar{\mathbf{x}}_{ij} - \bar{\mathbf{x}}_{ir}, \qquad \bar{\mathbf{X}}'_j = \bar{\mathbf{X}}_j - \bar{\mathbf{X}}_r$$

as the relative image and space coordinates. We can immediately obtain a simplified affine projection equation in terms of relative coordinates.

$$\bar{\mathbf{x}}'_{ij} = \mathbf{A}_i \bar{\mathbf{X}}'_j \tag{2.5}$$

Actually, the translation term $\bar{\mathbf{T}}_i$ is exactly the image of world origin. It is easy to verify that the centroid of a set of space points is projected to the centroid of their images. In practice, we can simply choose the centroid as the reference point, then the translation term vanishes if all the image points in each frame are registered to the corresponding centroid. The affine matrix $\mathbf{A}_i$ has six independent variables which encapsulate both intrinsic and extrinsic parameters of the affine camera. According to RQ decomposition [3], matrix $\mathbf{A}_i$ can be uniquely decomposed into the following form.

$$\mathbf{A}_i = \mathbf{K}_{Ai}\mathbf{R}_{Ai} = \begin{bmatrix} \alpha_{1i} & \zeta_i \\ & \alpha_{2i} \end{bmatrix} \begin{bmatrix} \mathbf{r}_{1i}^T \\ \mathbf{r}_{2i}^T \end{bmatrix} \tag{2.6}$$

where $\mathbf{K}_{Ai}$ is the intrinsic calibration matrix. In accordance to the camera matrix of perspective projection, α_{1i} and α_{2i} are the scaling factors of the two image axes and α_{1i}/α_{2i} is defined as the aspect ratio, ζ_i is the skew factor of the affine camera. For most CCD cameras, we usually assume unit aspect ratio $\alpha_{1i} = \alpha_{2i} = \alpha_i$, and zero skew $\zeta_i = 0$. $\mathbf{R}_{Ai}$ is the rotation matrix, $\mathbf{r}_{1i}^T$ and $\mathbf{r}_{2i}^T$ are the first two rows of the rotation matrix with the constraint

$$\mathbf{r}_{1i}^T \mathbf{r}_{2i} = 0, \qquad \|\mathbf{r}_{1i}\|^2 = \|\mathbf{r}_{2i}\|^2 = 1 \tag{2.7}$$

while the third row of the rotation matrix can always be recovered as $\mathbf{r}_{3i} = \mathbf{r}_{1i} \times \mathbf{r}_{2i}$. From the above analysis, we can easily see that the affine matrix $\mathbf{A}_i$ has six degrees of freedom. Under affine assumption, the camera projection is usually modeled by three special cases, i.e. orthographic projection, weak perspective projection, and para-perspective projection, as shown in Fig. 2.1.

Orthographic projection is the most simple approximation. In this case, it is assumed $\alpha_{1i} = \alpha_{1i} = 1$ and $\zeta_i = 0$. Thus the projection can be modelled as

$$\mathbf{K}_{ortho} = \begin{bmatrix} 1 & 0 \\ 0 & 1 \end{bmatrix} \tag{2.8}$$

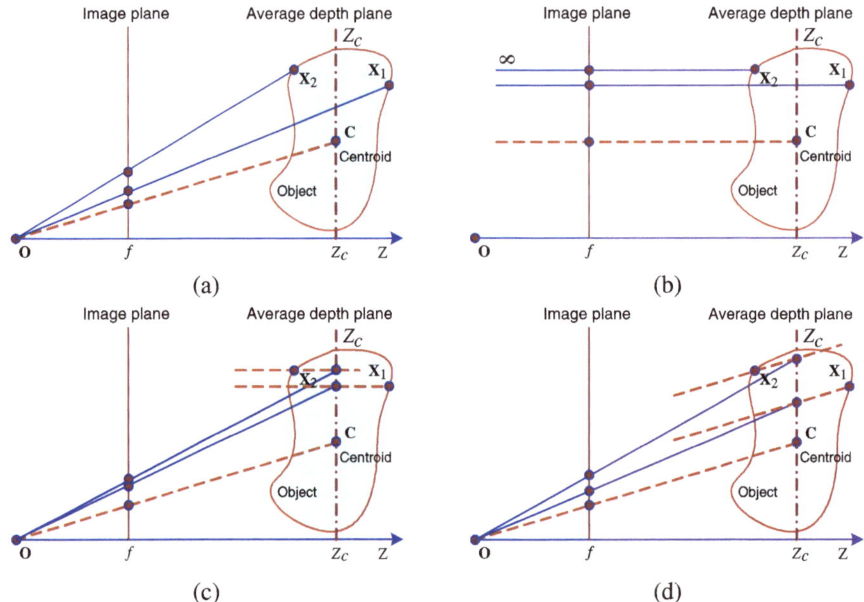

Fig. 2.1 The imaging process of different projection models. (**a**) Perspective projection; (**b**) Orthographic projection; (**c**) Weak-perspective projection; (**d**) Para-perspective projection. **O** is the optical center, $Z = f$ is the image plane. **C** is the centroid of the object, $Z = Z_c$ is the average depth plane, $\mathbf{X}_1$ and $\mathbf{X}_2$ are two space points on the object

where the subscript index i is omitted. In a weak-perspective projection, the space point is first projected to the average depth plane via orthographic projection, then projected to the image by perspective projection. Thus, the scaling factor is included as $\alpha_{1i} = \alpha_{2i} = \alpha_i$ and $\zeta_i = 0$, which is equivalent to a scaled orthography.

$$\mathbf{K}_{weak} = \alpha \mathbf{K}_{ortho} = \alpha \begin{bmatrix} 1 & 0 \\ 0 & 1 \end{bmatrix} \tag{2.9}$$

Thus, the weak-perspective projection can model the scaling effect caused by depth changes between images. It is suitable for objects with small depth variations. Para-perspective is a more generalized affine model which is a step closer to perspective projection. As shown in Fig. 2.1, the main difference between para-perspective and weak-perspective projection is that the space point is first projected to the average depth plane along the line passing through optical center and the centroid of the object. Thus it not only models the scaling of weak perspective, but also the apparent result of an object moving towards the edge of the image. Please refer to [6, 8] for more details on para-perspective projection. It can be verified that weak-perspective is a zero-order approximation of full perspective projection, while paraperspective is a first-order approximation.

2.3 Quasi-Perspective Projection Model

In this section, we will propose a new quasi-perspective projection model to fill the gap between simplicity of affine camera and accuracy of perspective projection.

2.3.1 Quasi-Perspective Projection

Under perspective projection, the image formation process is shown in Fig. 2.2. In order to ensure that large overlapping part of the object is reconstructed, the camera usually undergoes really small movements across adjacent views, especially for images of a video sequence.

Suppose $O_w - X_w Y_w Z_w$ is a world coordinate system selected on the object to be reconstructed. $O_i - X_i Y_i Z_i$ is the camera coordinate system with O_i being the optical center of the camera. Without loss of generality, we assume that there is a reference camera system $O_r - X_r Y_r Z_r$. Since the world system can be set freely, we align it with the reference frame as illustrated in Fig. 2.2. Therefore, the rotation $\mathbf{R}_i$ of frame i with respect to the reference frame is the same as the rotation of the camera to the world system.

Definition 2.1 (Axial and lateral rotation) The orientation of a camera is usually described by roll-pitch-yaw angles. For the i-th frame, we define the pitch, yaw, and roll as the rotations α_i, β_i, and γ_i of the camera with respect to the X_w, Y_w, and Z_w axes of the world system. As shown in Fig. 2.2, the optical axis of the cameras usually point towards the object. For convenience of discussion, we define γ_i as the axial rotation angle, and define α_i and β_i as lateral rotation angles.

Proposition 2.1 *Suppose the camera undergoes small lateral rotation with respect to the reference frame, then the variation of projective depth λ_{ij} is mainly proportional to the*

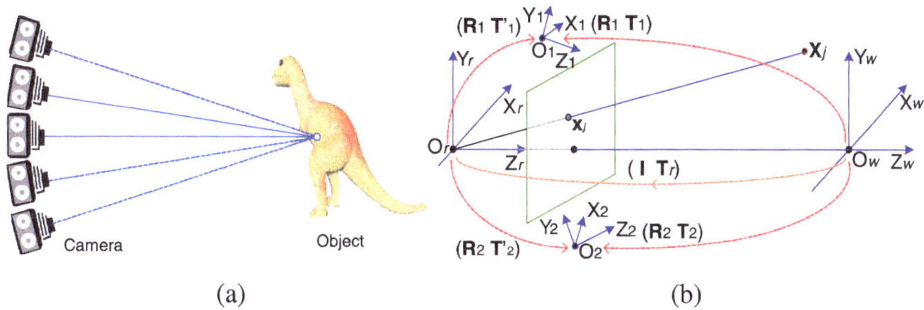

Fig. 2.2 Imaging process of camera take image sequence around an object. (**a**) Camera setup with respect to the object; (**b**) The relationship of world coordinate system and camera systems at different viewpoint

depth of the space point. The projective depths of a point at different views have similar trend of variation.

Proof Suppose the rotation matrix and translation vector of the ith frame with respect to the world system are

$$\mathbf{R}_i = \begin{bmatrix} \mathbf{r}_{1i}^T \\ \mathbf{r}_{2i}^T \\ \mathbf{r}_{3i}^T \end{bmatrix}, \qquad \mathbf{T}_i = \begin{bmatrix} t_{xi} \\ t_{yi} \\ t_{zi} \end{bmatrix} \tag{2.10}$$

Then, the projection matrix can be written as

$$\begin{aligned} \mathbf{P}_i &= \mathbf{K}_i[\mathbf{R}_i, \mathbf{T}_i] \\ &= \begin{bmatrix} f_i\mathbf{r}_{1i}^T + \varsigma_i\mathbf{r}_{2i}^T + u_{0i}\mathbf{r}_{3i}^T & f_i t_{xi} + \varsigma_i t_{yi} + u_{0i} t_{zi} \\ \kappa_i f_i \mathbf{r}_{2i}^T + v_{0i}\mathbf{r}_{3i}^T & \kappa_i f_i t_{yi} + v_{0i} t_{zi} \\ \mathbf{r}_{3i}^T & t_{zi} \end{bmatrix} \end{aligned} \tag{2.11}$$

The rotation matrix can be decomposed into rotations around the three axes of the world frame.

$$\begin{aligned} \mathbf{R}_i &= \mathbf{R}(\gamma_i)\mathbf{R}(\beta_i)\mathbf{R}(\alpha_i) \\ &= \begin{bmatrix} \cos\gamma_i, & -\sin\gamma_i, & 0 \\ \sin\gamma_i, & \cos\gamma_i, & 0 \\ 0, & 0, & 1 \end{bmatrix} \begin{bmatrix} \cos\beta_i, & 0, & \sin\beta_i \\ 0, & 1, & 0 \\ -\sin\beta_i, & 0, & \cos\beta_i \end{bmatrix} \begin{bmatrix} 1, & 0, & 0 \\ 0, & \cos\alpha_i, & -\sin\alpha_i \\ 0, & \sin\alpha_i, & \cos\alpha_i \end{bmatrix} \\ &= \begin{bmatrix} \cos\gamma_i\cos\beta_i, & \cos\gamma_i\sin\beta_i\sin\alpha_i - \sin\gamma_i\cos\alpha_i, & \cos\gamma_i\sin\beta_i\cos\alpha_i + \sin\gamma_i\sin\alpha_i \\ \sin\gamma_i\cos\beta_i, & \sin\gamma_i\sin\beta_i\sin\alpha_i + \cos\gamma_i\cos\alpha_i, & \sin\gamma_i\sin\beta_i\cos\alpha_i - \cos\gamma_i\sin\alpha_i \\ -\sin\beta_i, & \cos\beta_i\sin\alpha_i, & \cos\beta_i\cos\alpha_i \end{bmatrix} \end{aligned} \tag{2.12}$$

Inserting (2.11) and (2.12) into (2.1), we have

$$\lambda_{ij} = [\mathbf{r}_{3i}^T, t_{zi}]\mathbf{X}_j = -(\sin\beta_i)x_j + (\cos\beta_i\sin\alpha_i)y_j + (\cos\beta_i\cos\alpha_i)z_j + t_{zi} \tag{2.13}$$

From Fig. 2.2, we know that the rotation angles $\alpha_i, \beta_i, \gamma_i$ of the camera to the world system are the same as those to the reference frame. Under small lateral rotations, i.e., small angles of α_i and β_i, we have

$$\sin\beta_i \ll \cos\beta_i\cos\alpha_i, \qquad \cos\beta_i\sin\alpha_i \ll \cos\beta_i\cos\alpha_i \tag{2.14}$$

Thus, (2.13) can be approximated by

$$\lambda_{ij} \approx (\cos\beta_i\cos\alpha_i)z_j + t_{zi} \tag{2.15}$$

All features $\{x_{ij} | j = 1, \ldots, n\}$ in the ith frame correspond to the same rotation angles $\alpha_i, \beta_i, \gamma_i$, and translation t_{zi}. It is evident from (2.15) that the projective depths of a point in all frames have similar trend of variation, which are proportional to the value of z_j. Actually, the projective depths have no relation with the axial rotation γ_i. □

Proposition 2.2 *Under small lateral rotations and a further assumption that the distance from the camera to an object is significantly greater than the object depth, i.e.,* $t_{zi} \gg z_j$, *the ratio of* $\{\lambda_{ij} | i = 1, \ldots, m\}$ *corresponding to any two different frames can be approximated by a constant.*

Proof Let us take the reference frame as an example, the ratio of the projective depths of any frame i to those of the reference frame can be written as

$$\begin{aligned} \mu_i &= \frac{\lambda_{rj}}{\lambda_{ij}} \approx \frac{(\cos\beta_r \cos\alpha_r) z_j + t_{zr}}{(\cos\beta_i \cos\alpha_i) z_j + t_{zi}} \\ &= \frac{\cos\beta_r \cos\alpha_r (z_j / t_{zi}) + t_{zr}/t_{zi}}{\cos\beta_i \cos\alpha_i (z_j / t_{zi}) + 1} \end{aligned} \tag{2.16}$$

where $\cos\beta_i \cos\alpha_i \leq 1$. Under the assumption that $t_{zi} \gg z_j$, the ratio can be approximated by

$$\mu_i = \frac{\lambda_{rj}}{\lambda_{ij}} \approx \frac{t_{zr}}{t_{zi}} \tag{2.17}$$

All features in a frame have the same translation term. Therefore, we can see from (2.17) that the projective depth ratios of two frames for all features have the same approximation μ_i. □

According to Proposition 2.2, we have $\lambda_{ij} = \frac{1}{\mu_i}\lambda_{rj}$. Thus the perspective projection equation (2.1) can be approximated by

$$\frac{1}{\mu_i}\lambda_{rj}\mathbf{x}_{ij} = \mathbf{P}_i \mathbf{X}_j \tag{2.18}$$

Let us denote λ_{rj} as $\frac{1}{\ell_j}$, and reformulate (2.18) to

$$\mathbf{x}_{ij} = \mathbf{P}_{qi}\mathbf{X}_{qj} \tag{2.19}$$

where

$$\mathbf{P}_{qi} = \mu_i \mathbf{P}_i, \mathbf{X}_{qj} = \ell_j \mathbf{X}_j \tag{2.20}$$

We call (2.19) quasi-perspective projection model. Compared with general perspective projection, the quasi-perspective model assumes that projective depths between different frames are defined up to a constant μ_i. Thus, the projective depths are implicitly embedded in the scalars of the homogeneous structure $\mathbf{X}_{qj}$ and the projection matrix $\mathbf{P}_{qi}$, and the difficult problem of estimating the unknown depths is avoided. The model is more general than affine projection model (2.4), where all projective depths are simply assumed to be equal to $\lambda_{ij} = 1$.

2

2.3.2 Error Analysis of Different Models

In the following section, we will give a heuristic analysis on imaging errors of quasi-perspective and affine camera models with respect to the general perspective projection. For simplicity, the subscript 'i' of the frame number is omitted hereafter.

Suppose the intrinsic parameters of the cameras are known, and all images are normalized by the cameras as $\mathbf{K}_i^{-1}\mathbf{x}_{ij} \rightarrow \mathbf{x}_{ij}$. Then, the projection matrices under different projection model can be written as

$$\mathbf{P} = \begin{bmatrix} \mathbf{r}_1^T & t_x \\ \mathbf{r}_2^T & t_y \\ \mathbf{r}_3^T & t_z \end{bmatrix}, \quad \mathbf{r}_3^T = [-\sin\beta, \cos\beta\sin\alpha, \cos\beta\cos\alpha] \tag{2.21}$$

$$\mathbf{P}_q = \begin{bmatrix} \mathbf{r}_1^T & t_x \\ \mathbf{r}_2^T & t_y \\ \mathbf{r}_{3q}^T & t_z \end{bmatrix}, \quad \mathbf{r}_{3q}^T = [0, 0, \cos\beta\cos\alpha] \tag{2.22}$$

$$\mathbf{P}_a = \begin{bmatrix} \mathbf{r}_1^T & t_x \\ \mathbf{r}_2^T & t_y \\ \mathbf{0}^T & t_z \end{bmatrix}, \quad \mathbf{0}^T = [0, 0, 0] \tag{2.23}$$

where $\mathbf{P}$ is the projection matrix of perspective projection, $\mathbf{P}_q$ is that of quasi-perspective assumption, and $\mathbf{P}_a$ is that of affine projection. It is clear that the main difference between the projection matrices lies only in the last row. For a space point $\bar{\mathbf{X}} = [x, y, z]^T$, its projection under different camera models is given by

$$\mathbf{m} = \mathbf{P}\begin{bmatrix} \bar{\mathbf{X}} \\ 1 \end{bmatrix} = \begin{bmatrix} u \\ v \\ \mathbf{r}_3^T\bar{\mathbf{X}} + t_z \end{bmatrix} \tag{2.24}$$

$$\mathbf{m}_q = \mathbf{P}_q\begin{bmatrix} \bar{\mathbf{X}} \\ 1 \end{bmatrix} = \begin{bmatrix} u \\ v \\ \mathbf{r}_{3q}^T\bar{\mathbf{X}} + t_z \end{bmatrix} \tag{2.25}$$

$$\mathbf{m}_a = \mathbf{P}_a\begin{bmatrix} \bar{\mathbf{X}} \\ 1 \end{bmatrix} = \begin{bmatrix} u \\ v \\ t_z \end{bmatrix} \tag{2.26}$$

where

$$u = \mathbf{r}_1^T\bar{\mathbf{X}} + t_x, \; v = \mathbf{r}_2^T\bar{\mathbf{X}} + t_y \tag{2.27}$$

$$\mathbf{r}_3^T\bar{\mathbf{X}} = -(\sin\beta)x + (\cos\beta\sin\alpha)y + (\cos\beta\cos\alpha)z \tag{2.28}$$

$$\mathbf{r}_{3q}^T\bar{\mathbf{X}} = (\cos\beta\cos\alpha)z \tag{2.29}$$

and the nonhomogeneous image points can be denoted as

$$\bar{\mathbf{m}} = \frac{1}{\mathbf{r}_3^T \bar{\mathbf{X}} + t_z} \begin{bmatrix} u \\ v \end{bmatrix} \tag{2.30}$$

$$\bar{\mathbf{m}}_q = \frac{1}{\mathbf{r}_{3q}^T \bar{\mathbf{X}} + t_z} \begin{bmatrix} u \\ v \end{bmatrix} \tag{2.31}$$

$$\bar{\mathbf{m}}_a = \frac{1}{t_z} \begin{bmatrix} u \\ v \end{bmatrix} \tag{2.32}$$

The point $\bar{\mathbf{m}}$ is an ideal image by perspective projection. Let us define $\mathbf{e}_q = |\bar{\mathbf{m}}_q - \bar{\mathbf{m}}|$ as the error of quasi-perspective, and $\mathbf{e}_a = |\bar{\mathbf{m}}_a - \bar{\mathbf{m}}|$ as the error of affine, where '$|\cdot|$' stands for the norm of a vector. Then, we have

$$\begin{aligned} \mathbf{e}_q &= |\bar{\mathbf{m}}_q - \bar{\mathbf{m}}| \\ &= \left| \frac{\mathbf{r}_3^T \bar{\mathbf{X}} + t_z}{\mathbf{r}_{3q}^T \bar{\mathbf{X}} + t_z} \bar{\mathbf{m}} - \bar{\mathbf{m}} \right| = \det\left(\frac{(\mathbf{r}_3^T - \mathbf{r}_{3q}^T)\bar{\mathbf{X}}}{\mathbf{r}_{3q}^T \bar{\mathbf{X}} + t_z} \right) |\bar{\mathbf{m}}| \\ &= \det\left(\frac{-(\sin\beta)x + (\cos\beta\sin\alpha)y}{(\cos\beta\cos\alpha)z + t_z} \right) |\bar{\mathbf{m}}| \end{aligned} \tag{2.33}$$

$$\begin{aligned} \mathbf{e}_a &= |\bar{\mathbf{m}}_a - \bar{\mathbf{m}}| \\ &= \left| \frac{\mathbf{r}_3^T \bar{\mathbf{X}} + t_z}{t_z} \bar{\mathbf{m}} - \bar{\mathbf{m}} \right| = \det\left(\frac{\mathbf{r}_3^T \bar{\mathbf{X}}}{t_z} \right) |\bar{\mathbf{m}}| \\ &= \det\left(\frac{-(\sin\beta)x + (\cos\beta\sin\alpha)y + (\cos\beta\cos\alpha)z}{t_z} \right) |\bar{\mathbf{m}}| \end{aligned} \tag{2.34}$$

Based on the above equations, it is rational to state the following results for different projection models.

1. The axial rotation angle γ around Z-axis has no influence on the images of $\bar{\mathbf{m}}$, $\bar{\mathbf{m}}_q$ and $\bar{\mathbf{m}}_a$.
2. When the distance of a camera to an object is much larger than the object depth, both $\bar{\mathbf{m}}_q$ and $\bar{\mathbf{m}}_a$ are close to $\bar{\mathbf{m}}$.
3. When the camera system is aligned with the world system, i.e., $\alpha = \beta = 0$, we have $\mathbf{r}_{3q}^T = \mathbf{r}_3^T = [0, 0, 1]$ and $\mathbf{e}_q = 0$. Thus $\bar{\mathbf{m}}_q = \bar{\mathbf{m}}$, and the quasi-perspective assumption is equivalent to perspective projection.
4. When the rotation angles α and β are small, we have $\mathbf{e}_q < \mathbf{e}_a$, i.e., the quasi-perspective assumption is more accurate than affine assumption.
5. When the space point lies on the plane through the world origin and perpendicular to the principal axis, i.e., the direction of $\mathbf{r}_3^T$, we have $\alpha = \beta = 0$ and $z = 0$. It is easy to verify that $\bar{\mathbf{m}} = \bar{\mathbf{m}}_q = \bar{\mathbf{m}}_a$.

2.4 Experimental Evaluations

During simulation, we randomly generated 200 points within a cube of $20 \times 20 \times 20$ in space as shown in Fig. 2.3(a), only the first 50 points are displayed for simplicity. The depth variation in Z-direction of the space points is shown in Fig. 2.3(b). We simulated 10 images from these points by perspective projection. The image size is set at 800×800. The camera parameters are set as follows: focal lengths are set randomly between 900 and 1100, the principal point is set at the image center, and the skew is zero. The rotation angles are set randomly between $\pm 5°$. The X and Y positions of the cameras are set randomly between ± 15, while the Z position is spaced evenly from 200 to 220. The true projective depths λ_{ij} associated with these points across 10 different views are shown in Fig. 2.3(c), where the values are given after normalization so that they have unit mean.

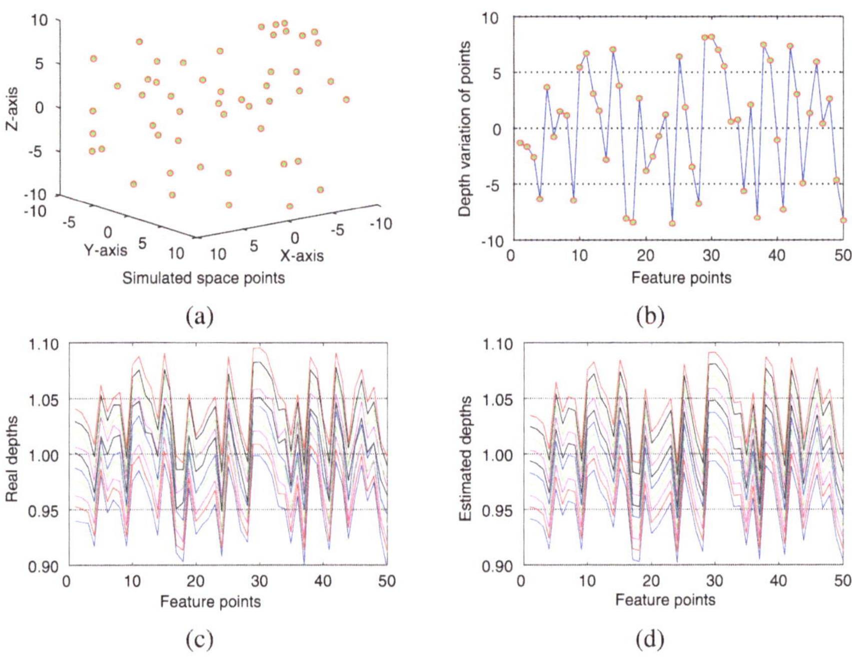

Fig. 2.3 Evaluation on projective depth approximation of the first 50 points. (**a**) Coordinates and distribution of the synthetic space points; (**b**) The depth variation of the space points; (**c**) The real projective depths of the imaged points after normalization; (**d**) The approximated projective depths under quasi-perspective assumption

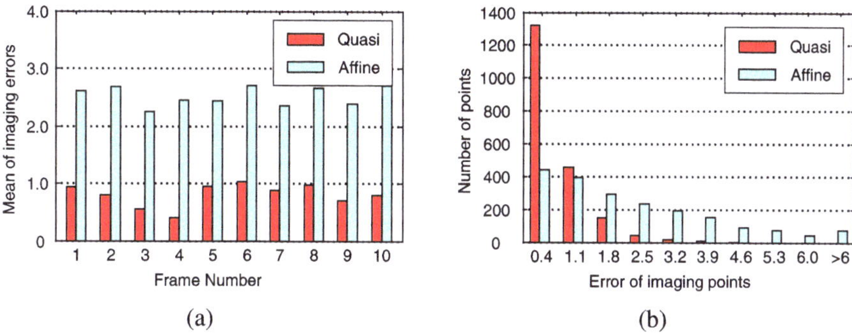

Fig. 2.4 Evaluation of the imaging errors by different camera models. (**a**) The mean error of the generated images by the quasi-perspective and affine projection models; (**b**) The histogram distribution of the errors by different projection models

2.4.1 Imaging Errors

Using the simulated data, we estimate λ_{1j} and μ_i from (2.15) and (2.16), and construct the estimated projective depths from $\hat{\lambda}_{ij} = \frac{\lambda_{1j}}{\mu_i}$. The normalized result is shown in Fig. 2.3(d). We can see from experiment that the recovered projective depths are very close to the ground truths. They are generally in proportion to the depth variation of space points in the Z-direction. If we adopt affine camera model, it is equivalent to setting all projective depths to $\lambda_{ij} = 1$. The error is obviously much bigger than that of the quasi-perspective assumption.

According to projection equations (2.30) to (2.34), different images will be obtained if we adopt different camera models. We generated three sets of images using the simulated space points via general perspective projection model, affine camera model, and quasi-perspective projection model. We compared the errors of quasi-perspective projection model (2.33) and affine assumption (2.34). The mean error of different models in each frame is shown in Fig. 2.4(a), the histogram distribution of the errors for all 200 points across 10 frames is shown in Fig. 2.4(b). Results indicate that the error of quasi-perspective assumption is much smaller than that under affine assumption.

2.4.2 Influence of Imaging Conditions

The proposed quasi-perspective model is based on the assumption of small camera movement. We investigated the influence of different imaging conditions to the model. Initially, we fix the camera position as given in the first test and vary the amplitude of rotation angles from $\pm 5°$ to $\pm 50°$ in steps of $5°$. At each step, we check the relative error of the recovered

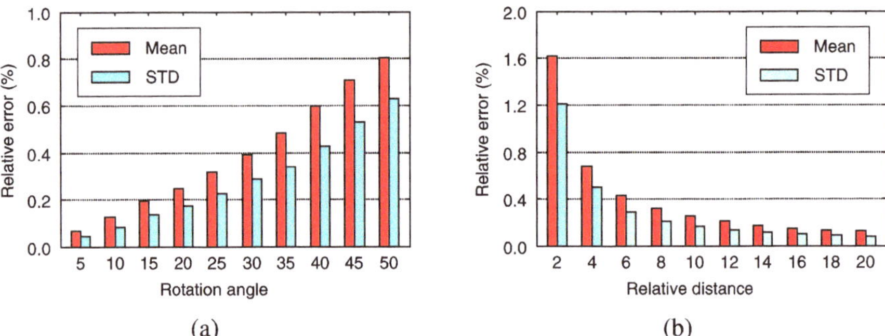

Fig. 2.5 Evaluation on quasi-perspective projection under different imaging conditions. (**a**) The relative error of the estimated depths under different rotation angles; (**b**) The relative error with respect to different relative distances

projective depths, which is defined as

$$e_{ij} = \frac{|\lambda_{ij} - \hat{\lambda}_{ij}|}{\lambda_{ij}} \times 100\,(\%) \tag{2.35}$$

where $\hat{\lambda}_{ij}$ is the estimated projective depth. We carried out 100 independent tests at each step so as to obtain a statistically meaningful result. The mean and standard deviation of e_{ij} are shown in Fig. 2.5(a).

Then, we fix the rotation angles at $\pm 5^\circ$ and vary the relative distance of a camera to an object (i.e. the ratio between the distance of a camera to an object center and the object depth) from 2 to 20 in steps of 2. The mean and standard deviation of e_{ij} at each step for 100 tests are shown in Fig. 2.5(b). Result shows that the quasi-perspective projection is a good approximation ($e_{ij} < 0.5\%$) when the rotation angles are less than $\pm 35^\circ$ and the relative distance is larger than 6. Please note that the result is obtained from noise free data.

2.5 Closure Remarks

2.5.1 Conclusion

In this chapter, we proposed a quasi-perspective projection model and analyzed the projection errors of different projection models. The proposed model is a trade-off between affine and perspective projection. It is computationally simple with better accuracy than affine approximation. The proposed model is suitable for structure and motion factorization of a short sequence with small camera motions. It should be noted that the small rotation assumption of the proposed model is not a limiting factor and is usually satisfied in many real world applications. During image acquisition of an object to be reconstructed, we tend to control the camera movement so as to guarantee large overlapping part, which also facilitates the feature tracking process. Some geometrical properties of the model in one view and two-view [11] will be presented in the next chapter. The application details to structure and motion factorization [12] will be given in Chap. 9.

2.5.2 Review Questions

1. *Affine camera model.* Provide the general form of the projection matrix under affine camera model. Show that the translation term corresponds to the image of world origin, and it can be removed if the image points are registered to the corresponding centroid. Show that an infinite point in space will also be projected to an infinite point in the image. Illustrate the imaging process of orthographic projection and weak-perspective projection.
2. *Quasi-perspective projection.* Elaborate on the two major assumptions of the quasi-perspective projection. Derive the projection matrix under quasi-perspective projection. What is the main difference between affine and quasi-perspective projection? Present the projection error of the two approximations and make a comparison.

References

1. Aloimonos, J.Y.: Perspective approximation. Image Vis. Comput. **8**(3), 177–192 (1990)
2. Faugeras, O.: Three-Dimensional Computer Vision: A Geometric Viewpoint. MIT Press, Cambridge (1993)
3. Hartley, R.I., Zisserman, A.: Multiple View Geometry in Computer Vision, 2nd edn. Cambridge University Press, Cambridge (2004). ISBN: 0521540518
4. Kanatani, K., Sugaya, Y., Ackermann, H.: Uncalibrated factorization using a variable symmetric affine camera. IEICE Trans. Inf. Syst. **E90-D**(5), 851–858 (2007)
5. Mundy, J.L., Zisserman, A.: Geometric Invariance in Computer Vision. MIT Press, Cambridge (1992)
6. Ohta, Y.I., Maenobu, K., Sakai, T.: Obtaining surface orientation from texels under perspective projection. In: Proc. International Joint Conferences on Artificial Intelligence, pp. 746–751 (1981)
7. Oliensis, J., Hartley, R.: Iterative extensions of the Sturm/Triggs algorithm: Convergence and nonconvergence. IEEE Trans. Pattern Anal. Mach. Intell. **29**(12), 2217–2233 (2007)
8. Poelman, C., Kanade, T.: A paraperspective factorization method for shape and motion recovery. IEEE Trans. Pattern Anal. Mach. Intell. **19**(3), 206–218 (1997)
9. Shapiro, L.S., Zisserman, A., Brady, M.: 3D motion recovery via affine epipolar geometry. Int. J. Comput. Vis. **16**(2), 147–182 (1995)
10. Wang, G., Wu, J.: Quasi-perspective projection with applications to 3D factorization from uncalibrated image sequences. In: Proc. of IEEE Conference on Computer Vision and Pattern Recognition, pp. 1–8 (2008)
11. Wang, G., Wu, J.: The quasi-perspective model: Geometric properties and 3D reconstruction. Pattern Recogn. **43**(5), 1932–1942 (2010)
12. Wang, G., Wu, J.: Quasi-perspective projection model: Theory and application to structure and motion factorization from uncalibrated image sequences. Int. J. Comput. Vis. **87**(3), 213–234 (2010)

3 Geometrical Properties of Quasi-Perspective Projection

Abstract The chapter investigates geometrical properties of quasi-perspective projection model in one and two-view geometry. The main results are as follows. (i) Quasi-perspective projection matrix has nine degrees of freedom, and the parallelism along X and Y directions in world system are preserved in images. (ii) Quasi-fundamental matrix can be simplified to a special form with only six degrees of freedom. The fundamental matrix is invariant to any non-singular projective transformation. (iii) Plane induced homography under quasi-perspective model can be simplified to a special form defined by six degrees of freedom. The quasi-homography may be recovered from two pairs of corresponding points with known fundamental matrix. (iv) Any two reconstructions in quasi-perspective space are defined up to a non-singular quasi-perspective transformation.

Euclid taught me that without assumptions there is no proof.
Therefore, in any argument, examine the assumptions.

Eric Temple Bell (1883–1960)

3.1 Introduction

Recovering three-dimensional information from stereo views is a fundamental problem in computer vision and significant progress have been made during the last two decades. The most typical algorithm is stereo vision obtained from two images based on epipolar geometry [7]. For an image sequence, we usually adopt factorization based algorithm [14] to recover structure and motion parameters.

All structure from motion algorithms are based on certain assumptions of camera model. The most accurate model is based on perspective projection, which is complicated due to its nonlinearity [12, 19]. Some popular approximations include orthographic, weak perspective, and paraperspective projection models. They are generalized as affine camera [11, 15]. More recently, Wang and Wu [18] proposed a quasi-perspective projection model to fill the gap between simplicity of affine camera and accuracy of perspective projection.

G. Wang, Q.M.J. Wu, *Guide to Three Dimensional Structure and Motion Factorization*, Advances in Pattern Recognition,
DOI 10.1007/978-0-85729-046-5_3,

Fundamental matrix estimation is a central problem in stereo vision as it encapsulates the underlying epipolar geometry between images. Classical linear techniques for fundamental matrix estimation are sensitive to noise. Hartley [6] analyzed the problem and proposed a normalized eight-point algorithm to improve the stability and accuracy of computation. Zhang and Kanade [26] provided a good review on fundamental matrix estimation and uncertainty analysis. Hu *et al.* [8] proposed to use evolutionary agents for epipolar geometry estimation. Random sample consensus (RANSAC) paradigm [4] was originated for robust parameter estimation in presence of outliers that severely affect least-squares based techniques. Torr *et al.* [13] proposed to adopt RANSAC to estimate fundamental matrix. Cheng and Lai [1] proposed a consensus sampling technique to increase the probability of sampling inliers. Dellaert *et al.* [2] also proposed a robust method to reject outliers and reconstruct 3D scene geometry.

The concepts of affine camera and affine fundamental matrix are well established in [11, 15] as a generalization of orthographic, weak perspective, and paraperspective projections. Zhang and Xu [27] presented a general expression of fundamental matrix for both projective and affine cameras. Wolf and Shashua [23] investigated the recovery of affine fundamental matrix and structure of multiple planes moving relatively to each other under pure translation between two cameras. Mendonca and Cipolla [10] investigated the trifocal tensor for an affine trinocular rig. Guilbert *et al.* [5] presented a batch algorithm for recovering Euclidean camera motion from sparse data for affine camera. Lehmann *et al.* [9] proposed an integral projection approach to determine affine fundamental matrix directly from two sets of features without any correspondence or explicit constraint on the data. Shimshoni *et al.* [16] Presented a geometric interpretation for weak-perspective motion from two and three images. Zhang *et al.* [25] investigated the problem of structure reconstruction from the combination of perspective and weak-perspective images.

The quasi-perspective projection model was originally proposed for factorization based structure recovery from image sequences [21]. In this chapter, we will further investigate some geometrical properties of the model in one and two-view geometry [20]. Results are obtained in an analogous manner to that of full perspective and affine camera model.

The remaining part of the chapter is organized as follows. The property of quasi-perspective projection matrix is given in Sect. 3.2. The two-view geometry of the model is elaborated in Sect. 3.3. Some properties on quasi-perspective reconstruction are presented in Sect. 3.4. Extensive experimental evaluations on synthetic and real images are reported in Sects. 3.5 and 3.6 respectively.

3.2 One-View Geometrical Property

In Chap. 2, under the assumption that the camera is far away from the object with small lateral rotations and translation, we proved that the variation of projective depth λ_{ij} is mainly proportional to the depth of the associated space point, and the projective depths between different frames can be defined up to a constant. Thus the projective depths may be implicitly embedded in the scalars of the homogeneous structure and the projection

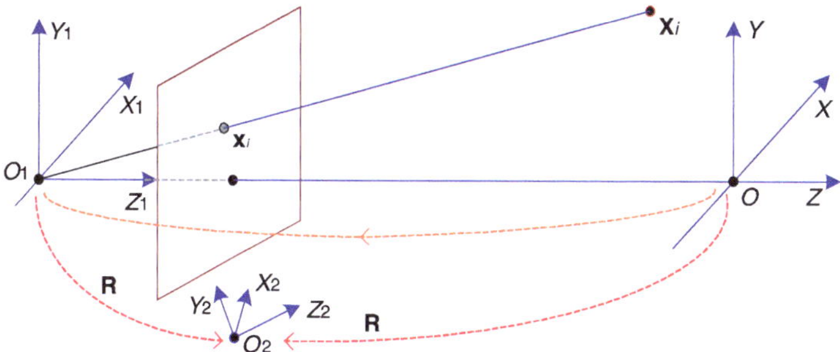

Fig. 3.1 The imaging process and relationship of the world coordinate system with respect to the camera system at different positions

matrix. Consequently, the imaging process is simplified to the following quasi-perspective projection.

$$\mathbf{x}_i = \mathbf{P}_q \mathbf{X}_{qi} \tag{3.1}$$

where $\mathbf{P}_q$ is the quasi-perspective projection matrix; $\mathbf{X}_{qi}$ is scale weighted space point in homogeneous form. In this section, we will present some geometrical properties of the model.

Result 3.1 *The quasi-perspective projection matrix has* 9 *degrees of freedom, and its last row is of the form* $\mathbf{P}_3^T \simeq [0, 0, *, *]$, *where '∗' stands for a nonzero entry.*

Proof As shown in Fig. 3.1, let us take the coordinate system $O_1 - X_1Y_1Z_1$ of first camera as a reference frame. Without loss of generality, we assume that the world coordinate system $O - XYZ$ is aligned with the reference system. Then the camera rotation $\mathbf{R}$ with respect to the reference frame is the same as the rotation with respect to the world system. Suppose the rotation and translation of the second view are $\mathbf{R} = [\mathbf{r}_1, \mathbf{r}_2, \mathbf{r}_3]^T = \mathbf{R}(\gamma)\mathbf{R}(\beta)\mathbf{R}(\alpha)$ and $\mathbf{T} = [t_1, t_2, t_3]^T$, where α, β, and γ are the rotation angles along the three axes X, Y, and Z respectively. Then, the last row of the projection matrix can be written as

$$\mathbf{P}_3^T = [\mathbf{r}_3^T, t_3] = [-\sin\beta, \cos\beta\sin\alpha, \cos\beta\cos\alpha, t_3] \tag{3.2}$$

Under quasi-perspective assumption of small rotations, we have

$$\{\sin\alpha, \sin\beta, \sin\gamma\} \to 0, \qquad \{\cos\alpha, \cos\beta, \cos\gamma\} \to 1 \tag{3.3}$$

which results to $\sin\beta \ll \cos\beta\cos\alpha \leq 1$, and $\cos\beta\sin\alpha \ll \cos\beta\cos\alpha \leq 1$. Thus, the last row (3.2) can be approximated by

$$\mathbf{P}_3^T \simeq [0, 0, \cos\beta\cos\alpha, t_z] = [0, 0, *, *] \tag{3.4}$$

As a consequence, the quasi-perspective projection matrix $\mathbf{P}_q$ has only 10 nonzero entries and 9 degrees of freedom since it is defined up to a global scale. The matrix $\mathbf{P}_q$ can

3

be linearly solved from a minimum of $4\frac{1}{2}$ space to image correspondences. In comparison, at least $5\frac{1}{2}$ correspondences are required for the recovery of a full perspective projection matrix, and 4 pairs of matches to recover affine model. □

Result 3.2 *Under quasi-perspective projection, parallel lines along X and Y directions of world system are mapped to parallel lines in an image.*

Proof In projective geometry, parallel lines in 3D space appear to intersect at a point at infinity, which is also called ideal point in $\mathbb{P}^3$. The intersection of the parallel lines along X and Y directions can be written in canonical form as $\mathbf{V}_x = [1, 0, 0, 0]^T$, $\mathbf{V}_y = [0, 1, 0, 0]^T$. Then, their images are obtained as

$$\mathbf{v}_x = \mathbf{P}_q \mathbf{V}_x = \begin{bmatrix} * \\ * \\ 0 \end{bmatrix}, \qquad \mathbf{v}_y = \mathbf{P}_q \mathbf{V}_y = \begin{bmatrix} * \\ * \\ 0 \end{bmatrix} \tag{3.5}$$

Obviously, both $\mathbf{v}_x$ and $\mathbf{v}_y$ locate at infinity, thus the parallelism is preserved along the X and Y directions. □

In a similar way, we can verify that parallel lines on the $O - XY$ coordinate plane or any other parallel planes also project to parallel lines in the image. However, the parallel relationship is not preserved in Z direction. This is different with respect to affine camera model where the parallelism is invariant.

When the space points are coplanar, we assume that the plane is $Z = 0$ without loss of generality, then the quasi-perspective projection (3.1) is simplified to

$$\mathbf{x}_i \simeq \mathbf{H}_\pi \begin{bmatrix} X_i \\ Y_i \\ 1 \end{bmatrix} = \begin{bmatrix} h_{11} & h_{12} & h_{13} \\ h_{21} & h_{22} & h_{23} \\ 0 & 0 & h_{33} \end{bmatrix} \begin{bmatrix} X_i \\ Y_i \\ 1 \end{bmatrix} \tag{3.6}$$

where $\mathbf{H}_\pi \in \mathbb{R}^{3\times 3}$ is called homography or perspectivity. There are 6 degrees of freedom in the perspectivity, so $\mathbf{H}_\pi$ can be recovered from 3 non-collinear space points with known positions. The form of the perspectivity in (3.6) is the same as that under affine model. While a general homography under perspective model has 8 degrees of freedom and at least 4 points are required for computation.

3.3 Two-View Geometrical Property

In this section, more geometrical properties of quasi-perspective projection are introduced in the context of two views.

3.3.1 Fundamental Matrix

Epipolar geometry is the intrinsic projective geometry between a pair of stereo images. As discussed in Chap. 1, the intrinsic geometry between two images can be encapsulated as

$$\mathbf{x}'^T\mathbf{F}\mathbf{x} = 0 \tag{3.7}$$

where $\mathbf{F} \in \mathbb{R}^{3\times3}$ is the so-called fundamental matrix; $\mathbf{x}$ and $\mathbf{x}'$ are a pair of corresponding points between the images. If the camera parameters are calibrated and the images are normalized as $\mathbf{y} = \mathbf{K}^{-1}\mathbf{x}$, $\mathbf{y}' = \mathbf{K}'^{-1}\mathbf{x}'$, then we have an analogous relation as $\mathbf{y}'^T\mathbf{E}\mathbf{y} = 0$, where $\mathbf{E}$ is named as essential matrix. The two matrices are related by

$$\mathbf{E}_q = \mathbf{K}'^T\mathbf{F}_q\mathbf{K} \tag{3.8}$$

Both the fundamental matrix and the essential matrix are rank-2 homogeneous matrices defined up to scales, thus they have only 7 degrees of freedom.

Result 3.3 *Both the fundamental matrix and the essential matrix under quasi-perspective projection can be simplified to the form of*

$$\begin{bmatrix} 0 & * & * \\ * & 0 & * \\ * & * & * \end{bmatrix}$$

which is defined by 5 degrees of freedom.

Proof Given the rotation $\mathbf{R} = \mathbf{R}(\gamma)\mathbf{R}(\beta)\mathbf{R}(\alpha)$ and the translation $\mathbf{t} = [t_1, t_2, t_3]^T$ between two views, the essential matrix can be computed from

$$\mathbf{E}_q = [\mathbf{t}]_\times\mathbf{R} = \begin{bmatrix} e_{11} & e_{12} & e_{13} \\ e_{21} & e_{22} & e_{23} \\ e_{31} & e_{32} & e_{33} \end{bmatrix} \tag{3.9}$$

where

$$[\mathbf{t}]_\times = \begin{bmatrix} 0 & t_3 & -t_2 \\ -t_3 & 0 & t_1 \\ t_2 & -t_1 & 0 \end{bmatrix}$$

$$e_{11} = t_3 \sin\gamma \cos\beta + t_2 \sin\beta$$

$$e_{21} = -t_3 \cos\gamma \cos\beta - t_1 \sin\beta$$

$$e_{12} = t_3(\cos\gamma \cos\alpha + \sin\gamma \sin\beta \sin\alpha) - t_2 \cos\beta \sin\alpha$$

$$e_{22} = t_3(\sin\gamma \cos\alpha - \cos\gamma \sin\beta \sin\alpha) + t_1 \cos\beta \sin\alpha$$

3

Under the condition of (3.3), we can obtain that $\{e_{11}, e_{22}\} \to 0$. Therefore, the essential matrix is simplified to

$$\mathbf{E}_q = \begin{bmatrix} 0 & e_{12} & e_{13} \\ e_{21} & 0 & e_{23} \\ e_{31} & e_{32} & e_{33} \end{bmatrix} = \begin{bmatrix} 0 & * & * \\ * & 0 & * \\ * & * & * \end{bmatrix} \tag{3.10}$$

Suppose the camera parameters are fixed as

$$\mathbf{K} = \mathbf{K}' = \begin{bmatrix} f_1 & 0 & u_0 \\ 0 & f_2 & v_0 \\ 0 & 0 & 1 \end{bmatrix} \tag{3.11}$$

Then, the quasi-fundamental matrix can be obtained from

$$\mathbf{F}_q = \mathbf{K}'^{-T}\mathbf{E}_q\mathbf{K}^{-1} = \begin{bmatrix} e_{11}/f_1^2 & e_{12}/(f_1 f_2) & * \\ e_{21}/(f_1 f_2) & e_{22}/f_2^2 & * \\ * & * & * \end{bmatrix}$$

$$= \begin{bmatrix} 0 & f_{12} & f_{13} \\ f_{21} & 0 & f_{23} \\ f_{31} & f_{32} & f_{33} \end{bmatrix} = \begin{bmatrix} 0 & * & * \\ * & 0 & * \\ * & * & * \end{bmatrix} \tag{3.12}$$

Both $\mathbf{E}_q$ and $\mathbf{F}_q$ have the same form with 7 entries which is defined up to a scale. However, the two matrices have only 5 degrees of freedom, since there is an additional rank-2 constraint. □

To address the rank-2 constraint, the quasi-fundamental matrix may be parameterized as follows.

$$\mathbf{F}_q = \begin{bmatrix} 0 & f_{12} & k_2 f_{12} \\ f_{21} & 0 & k_1 f_{21} \\ f_{31} & f_{32} & k_1 f_{31} + k_2 f_{32} \end{bmatrix} \tag{3.13}$$

The quasi-essential matrix can also be parameterized in a similar way. There are only 6 entries in (3.13), and the fundamental matrix can be estimated via nonlinear iterations. Under the above parametrization, the epipole in the second image is given by $\mathbf{e}' = [-k_1, -k_2, 1]^T$. It should be noted that the parameterization (3.13) degenerates when $\mathbf{e}'$ lies at infinity or the camera undergoes pure translation. In these cases, it is easy to verify that $f_{12} = f_{21} = 0$. Thus the first two columns of $\mathbf{F}_q$ are linearly dependent.

Result 3.4 *Given two quasi-perspective camera matrices* $\mathbf{P}_q$ *and* $\mathbf{P}'_q$, *the fundamental matrix between the two views can be recovered from* $\mathbf{F}_q = [\mathbf{e}']_\times \mathbf{P}'_q \mathbf{P}_q^+$, *where* $\mathbf{P}_q^+$ *denotes the pseudo-inverse of* $\mathbf{P}_q$. *The fundamental matrix is invariant to any non-singular projective transformation* $\mathbf{H} \in \mathbb{R}^{4\times 4}$. *i.e.* $\mathbf{F}_q$ *remains the same if we set* $\mathbf{P}_q \leftarrow \mathbf{P}_q\mathbf{H}$ *and* $\mathbf{P}'_q \leftarrow \mathbf{P}'_q\mathbf{H}$.

Proof Similar to the case of perspective projection as in [26], it is easy to obtain the following relationship.

$$\mathbf{F}_q = [\mathbf{e}']_\times \mathbf{P}'_q \mathbf{P}_q^+ \tag{3.14}$$

Suppose a space point $\mathbf{X}_i$ is projected to $\mathbf{x}_i$ and $\mathbf{x}'_i$ via the projection matrices $\mathbf{P}_q$ and $\mathbf{P}'_q$ respectively. If we apply a non-singular projective transformation $\mathbf{H}$ to the world system, i.e. $\mathbf{P}_q \leftarrow \mathbf{P}_q\mathbf{H}$, $\mathbf{P}'_q \leftarrow \mathbf{P}'_q\mathbf{H}$, and $\mathbf{X}_i \leftarrow \mathbf{H}^{-1}\mathbf{X}_i$. We can easily verify that the transformation does not change the images $\mathbf{x}_i$ and $\mathbf{x}'_i$. Thus the camera pairs $\{\mathbf{P}_q, \mathbf{P}'_q\}$ and $\{\mathbf{P}_q\mathbf{H}, \mathbf{P}'_q\mathbf{H}\}$ correspond to the same fundamental matrix as

$$\mathbf{F}_q = [\mathbf{e}']_\times \mathbf{P}'_q \mathbf{P}_q^+ = [\mathbf{e}']_\times (\mathbf{P}'_q\mathbf{H})(\mathbf{P}_q\mathbf{H})^+ \tag{3.15}$$

which indicates that the quasi-fundamental matrix is invariant to the transformation $\mathbf{H}$. Specifically, we can choose a certain transformation matrix to register the first camera to the world system and obtain the following projection matrices.

$$\mathbf{P}_q = \mathbf{K}[\,\mathbf{I}\,|\,0\,], \qquad \mathbf{P}'_q = \mathbf{K}'[\,\mathbf{R}\,|\,\mathbf{t}\,] \tag{3.16}$$

Then the epipole in the second image equals to $\mathbf{e}' = \mathbf{K}'\mathbf{t}$, and the fundamental matrix can be expressed by substituting (3.16) into (3.14).

$$\begin{aligned}\mathbf{F}_q &= [\mathbf{K}'\mathbf{t}]_\times \mathbf{K}'[\,\mathbf{R}\,|\,\mathbf{t}\,]\big(\mathbf{K}[\,\mathbf{I}\,|\,0\,]\big)^+ = \mathbf{K}'^{-T}\mathbf{t}_\times \mathbf{R}\mathbf{K}^{-1} \\ &= \mathbf{K}'^{-T}\mathbf{E}_q\mathbf{K}^{-1}\end{aligned} \tag{3.17}$$

Equation (3.17) derives (3.12) from a different viewpoint. □

Remark 3.1 For the computation of general fundamental matrix under perspective projection, we may adopt a normalized 8-point linear algorithm [6], iterative minimization algorithm of Sampson distance [22], 7-point nonlinear algorithm with rank-2 constraint [26], or the Gold Standard algorithm [7]. Please refer to [7, 26] for more details and a comparison of the above algorithms. Similarly, we have normalized 6-point linear algorithm and 5-point nonlinear algorithm for the estimation of quasi-perspective fundamental matrix. Usually, we adopt the linear algorithm for initial estimation, and utilize the Gold Standard algorithm to further optimize the fundamental matrix.

Remark 3.2 Under affine assumption, the optical center of an affine camera locates at infinity, it follows that all epipolar lines are parallel and both epipoles are at infinity. Thus the affine fundamental matrix is simplified to the form

$$\mathbf{F}_a = \begin{bmatrix} 0 & 0 & * \\ 0 & 0 & * \\ * & * & * \end{bmatrix}$$

which is already a rank-2 matrix with 4 degrees of freedom.

3

3.3.2 Plane Induced Homography

For coplanar space points, their images in two views are related with a planar homography, which is named as plane induced homography.

Result 3.5 *Under quasi-perspective projection, the plane induced homography can be simplified to the form of*

$$\mathbf{H}_q = \begin{bmatrix} * & * & * \\ * & * & * \\ 0 & 0 & * \end{bmatrix}$$

which has 6 *degrees of freedom.*

Proof Suppose $\mathbf{x}$ and $\mathbf{x}'$ are images of coplanar space point $\mathbf{X}$ in the two views, $\mathbf{H}_\pi$ and $\mathbf{H}'_\pi$ are the perspective homographies of the two views. Then from (3.6) we have

$$\mathbf{x} \simeq \mathbf{H}_\pi \mathbf{X}, \qquad \mathbf{x}' \simeq \mathbf{H}'_\pi \mathbf{X} \tag{3.18}$$

By eliminating $\mathbf{X}$ from (3.18), we have

$$\mathbf{x}' \simeq \mathbf{H}'_\pi \mathbf{H}_\pi^{-1} \mathbf{x} \tag{3.19}$$

where $\mathbf{H}_q = \mathbf{H}'_\pi \mathbf{H}_\pi^{-1}$ is called plane induced homography which can be expended as

$$\mathbf{H}_q = \mathbf{H}'_\pi \mathbf{H}_\pi^{-1} = \begin{bmatrix} h'_{11} & h'_{12} & h'_{13} \\ h'_{21} & h'_{22} & h'_{23} \\ 0 & 0 & h'_{33} \end{bmatrix} \begin{bmatrix} h_{11} & h_{12} & h_{13} \\ h_{21} & h_{22} & h_{23} \\ 0 & 0 & h_{33} \end{bmatrix}^{-1} = \begin{bmatrix} * & * & * \\ * & * & * \\ 0 & 0 & * \end{bmatrix} \tag{3.20}$$

The homography $\mathbf{H}_q$ is a full rank matrix with 6 degrees of freedom, and at least 3 non-collinear corresponding points can give a unique solution. □

It is easy to verify that the homography under affine camera model has the same form as (3.20). While general homography under perspective model has 8 degrees of freedom and at least 4 points are required for computation. A comparison of the entry numbers and degrees of freedom under different camera models of above discussed geometric matrices are tabulated in Table 3.1.

Table 3.1 The entry number and degrees of freedom (DOF) of different geometric matrices

Model		Projection matrix	Perspectivity matrix	Fundamental matrix	Essential matrix	Homography matrix
	Persp	12 (11)	9 (8)	9 (7)	9 (7)	9 (7)
Entry (DOF)	Quasi	10 (9)	7 (6)	7 (5)	7 (5)	7 (6)
	Affine	9 (8)	7 (6)	5 (4)	5 (4)	7 (6)

Result 3.6 *Given a fundamental matrix* $\mathbf{F}_q$*, the plane induced homography* $\mathbf{H}_q$ *may be recovered from two pairs of correspondences* $\mathbf{x}_i \leftrightarrow \mathbf{x}'_i, i = 1, 2$.

The result is obvious, since an additional correspondence of the epipoles $\mathbf{e} \leftrightarrow \mathbf{e}'$ can be obtained from the fundamental matrix as

$$\mathbf{F}_q \mathbf{e} = 0, \qquad \mathbf{F}_q^T \mathbf{e}' = 0 \tag{3.21}$$

Thus, if the two image points $\mathbf{x}_i, i = 1, 2$ are not collinear with the epipole $\mathbf{e}$, $\mathbf{H}_q$ can be uniquely determined from the three correspondences.

The homography $\mathbf{H}_\infty$ induced by the plane at infinity is called infinite homography. $\mathbf{H}_\infty$ is of great importance in stereo vision since it is closely related with camera calibration and affine reconstruction. According to the Result 3.6, the infinite homography may be computed from the correspondences of two vanishing points if the fundamental matrix is known. This is an interesting result for quasi-perspective projection model.

3.3.3 Computation with Outliers

In the above analysis, we assume all correspondences are inliers without mismatches. However, mismatches are inevitable in real application, and the result may be severely disturbed in presence of outliers. In this case, we usually adopt the RANSAC algorithm [4] to eliminate outliers and obtain a robust estimation. RANSAC algorithm is an iterative method to estimate parameters of a mathematical model and is computationally intensive. We will present a comparison on the number of trials required for different projection models.

Suppose the outlier-to-inlier ratio is $k = N_{outlier}/N_{inlier}$, the number of the minimum subset required to estimate the model is n. We want to ensure that at least one of the random samples is free from outliers with a probability of p. Then, the trial number N must satisfy

$$1 - p = \left(1 - \left(\frac{1}{k+1}\right)^n\right)^N \tag{3.22}$$

which leads to

$$N = \frac{\ln(1-p)}{\ln(1 - (\frac{1}{k+1})^n)} \tag{3.23}$$

Under the given probability p, the number of trials depend on the proportion k of outliers over inliers and the subset number n. In practice, we usually select a conservative probability $p = 0.99$. Table 3.2 shows the required number of trials under different conditions.

We can conclude from the table that the required number of trials increases sharply with an increase in subset number n and outlier ratio k. The quasi-perspective algorithm is

Table 3.2 The number of trials required for different models to ensure probability $p = 99\%$ with respect to different minimal subsets and outlier-to-inlier ratios

Model		Minimal subset	Outlier-to-inlier ratio					
			10%	20%	40%	60%	80%	100%
	Persp	8/7	8/7	18/15	66/47	196/122	506/280	1177/588
Fundamental	Quasi	6/5	6/5	12/9	33/23	75/46	155/85	293/146
	Affine	4	5	7	16	28	47	72
	Persp	4	5	7	16	28	47	72
Homography	Quasi	3	4	6	11	17	25	35
	Affine	3	4	6	11	17	25	35

computationally less intensive than perspective projection, especially for large proportion of outliers. As noted in Remark 3.1, we may adopt a normalized 8-point linear algorithm or 7-point nonlinear algorithm for fundamental estimation under perspective projection. Accordingly, we have 6-point linear algorithm and 5-point nonlinear algorithm for the computation of quasi-fundamental matrix. We can adopt the simple linear algorithm when the ratio k is small. However, it is wise to adopt a nonlinear algorithm for large outlier ratios so as to speed up computation.

3.4 3D Structure Reconstruction

Quasi-perspective projection is a special case of perspective projection, thus most theories on 3D reconstruction under perspective model may be applied directly to quasi-perspective model. Some important properties of quasi-perspective reconstruction are summarized as follows.

Result 3.7 *Under quasi-perspective assumption, a pair of canonical cameras can be defined as*

$$\mathbf{P}_q = [\,\mathbf{I}\,|\,\mathbf{0}\,], \qquad \mathbf{P}'_q = [\,\mathbf{M}_q\,|\,\mathbf{t}\,] \tag{3.24}$$

where $\mathbf{M}_q$ *is a* 3×3 *matrix with its last row of the form* $[\,0, 0, *\,]$.

Result 3.8 *Suppose* $(\mathbf{P}_{q1}, \mathbf{P}'_{q1}, \{\mathbf{X}_{1i}\})$ *and* $(\mathbf{P}_{q2}, \mathbf{P}'_{q2}, \{\mathbf{X}_{2i}\})$ *are two quasi-perspective reconstructions of a set of correspondences* $\mathbf{x}_i \leftrightarrow \mathbf{x}'_i$ *between two images. Then the two reconstructions are defined up to a quasi-perspective transformation as*

$$\mathbf{P}_{q2} = \mathbf{P}_{q1}\mathbf{H}_q, \qquad \mathbf{P}'_{q2} = \mathbf{P}'_{q1}\mathbf{H}_q, \qquad \mathbf{X}_{2i} = \mathbf{H}_q^{-1}\mathbf{X}_{1i}$$

where the transformation $\mathbf{H}_q$ *is a* 4×4 *non-singular matrix of the form*

$$\mathbf{H}_q = \begin{bmatrix} \mathbf{A}_{2\times 2} & \mathbf{B}_{2\times 2} \\ \mathbf{0}_{2\times 2} & \mathbf{C}_{2\times 2} \end{bmatrix} \tag{3.25}$$

Under quasi-perspective transformation $\mathbf{H}_q$, we have

$$\mathbf{P}_{q2}\mathbf{X}_{2i} = (\mathbf{P}_{q1}\mathbf{H}_q)(\mathbf{H}_q^{-1}\mathbf{X}_{1i}) = \mathbf{P}_{q1}\mathbf{X}_{1i} = \mathbf{x}_i \tag{3.26}$$

$$\mathbf{P}'_{q2}\mathbf{X}_{2i} = (\mathbf{P}'_{q1}\mathbf{H}_q)(\mathbf{H}_q^{-1}\mathbf{X}_{1i}) = \mathbf{P}'_{q1}\mathbf{X}_{1i} = \mathbf{x}'_i \tag{3.27}$$

It is easy to verify that the transformed camera matrices $\mathbf{P}_{q2} = \mathbf{P}_{q1}\mathbf{H}_q$ and $\mathbf{P}'_{q2} = \mathbf{P}'_{q1}\mathbf{H}_q$ remain the same form as quasi-perspective projection matrices as given in Result 3.1. The transformed space points can be written as

$$\mathbf{X}_{2i} = \mathbf{H}_q^{-1}\mathbf{X}_{1i} = \begin{bmatrix} \mathbf{A}^{-1} & -\mathbf{A}^{-1}\mathbf{B}\mathbf{C}^{-1} \\ \mathbf{0} & \mathbf{C}^{-1} \end{bmatrix} \mathbf{X}_{1i} \tag{3.28}$$

We observe that the parallelism along X and Y axes are preserved under the transformation $\mathbf{H}_q$, since ideal points $\mathbf{X}_{1i} = [1, 0, 0, 0]^T$ or $\mathbf{X}_{1i} = [0, 1, 0, 0]^T$ are mapped to ideal points according to (3.28).

For recovery of the 3D structure and camera motions, we may adopt a stratified reconstruction algorithm [3] to refine the structure from perspective to affine, and finally to the Euclidean space. In this chapter, we assume calibrated cameras. Therefore, we recover the metric structure directly from singular value decomposition (SVD) of the essential matrix [7]. The implementation of the reconstruction algorithm is summarized as follows.

1. Establish initial correspondences between the two images according to the method in [17];
2. Estimate quasi-fundamental matrix via RANSAC algorithm and eliminate outliers;
3. Optimize fundamental matrix via Gold Standard algorithm as stated in Remark 3.1 and recover the essential matrix from (3.8);
4. Perform SVD decomposition on the essential matrix and extract the camera projection matrices according to the method in [7], which will give four pairs of solutions;
5. Resolve ambiguity in the solution via trial and error. Take one pair of matching points as reference and reconstruct it from the above four solutions, only the true solution can make the reconstructed point lie in front of both cameras;
6. Compute 3D structure of all correspondences via triangulation from the recovered camera matrices;
7. Optimize the solution via bundle adjustment [7].

3.5 Evaluations on Synthetic Data

During simulation, we randomly generated 200 points within a cube of $20 \times 20 \times 20$ in space, and simulated two images from these points by perspective projection. The image

3

size is set at 800 × 800. The camera parameters are set as follows: focal lengths are set randomly between 1000 and 1100. Three rotation angles are set randomly between ±5°. The X and Y positions of the cameras are set randomly between ±15, while the Z positions are set randomly between 210 to 220. The synthetic imaging conditions are close to affine and quasi-perspective assumption.

3.5.1 Fundamental Matrix and Homography

We recovered the quasi-fundamental matrix $\mathbf{F}_q$ by normalized 6-point algorithm and calculated the epipolar residual error ε_{1i} which is defined as the distance of a point to the associated epipolar line.

$$\varepsilon_{1i} = \frac{1}{2}\left(dis(\mathbf{x}_i, \mathbf{F}_q^T \mathbf{x}_i')^2 + dis(\mathbf{x}_i', \mathbf{F}_q \mathbf{x}_i)^2\right) \tag{3.29}$$

where $dis(*, *)$ denotes the distance from a point to a line. The histogram distribution of the errors across all 200 correspondences is outlined in Fig. 3.2. Gaussian image noise was added to each image point during the test. As a comparison, we also recovered the general fundamental matrix $\mathbf{F}$ by normalized 8-point algorithm and the affine fundamental matrix $\mathbf{F}_a$ by normalized 4-point algorithm. We see that the error of quasi-perspective projection lies in between that of perspective projection and affine. Thus quasi-perspective fundamental matrix is a better approximation than affine fundamental matrix.

To evaluate the computation of homography, we set all space points on the plane $Z = 10$ and regenerated two images with the same camera parameters. Then, we recovered the plane induced homography $\mathbf{H}_q$ and $\mathbf{H}$ under quasi-perspective and perspective projection respectively, and evaluated the reprojection error as

$$\varepsilon_{2i} = \frac{1}{2}\left(d(\mathbf{x}_i, \mathbf{H}_q^{-1} \mathbf{x}_i')^2 + d(\mathbf{x}_i', \mathbf{H}_q \mathbf{x}_i)^2\right) \tag{3.30}$$

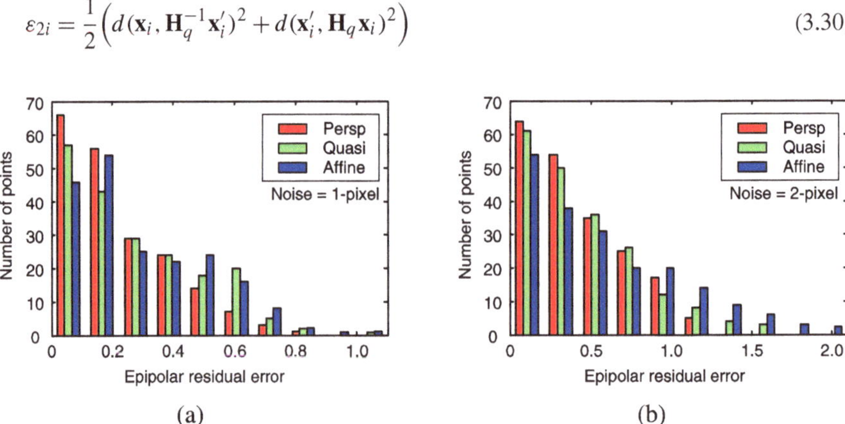

Fig. 3.2 Histogram distribution of the epipolar residual error under different camera models. (**a**) The result obtained with 1-pixel Gaussian noise; (**b**) The result obtained 2-pixel Gaussian noise

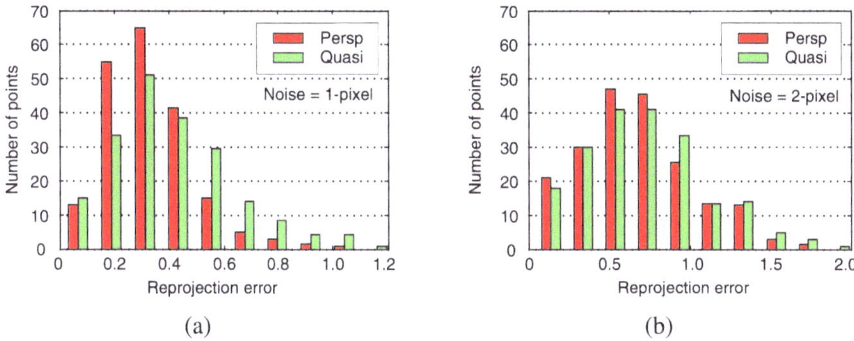

Fig. 3.3 Histogram distribution of reprojection errors by plane induced homography. (**a**) The result obtained with 1-pixel Gaussian noise; (**b**) The result obtained 2-pixel Gaussian noise

where $d(*, *)$ denotes the Euclidean distance between two feature points. The histogram distribution of the errors under different noise levels is shown in Fig. 3.3. It is clear that the error obtained from $\mathbf{H}_q$ is higher than that from $\mathbf{H}$. The homography under affine model is the same as $\mathbf{H}_q$ as noted in Sect. 3.3.2.

3.5.2 Outlier Removal

We randomly added 50 mismatches to the initial generated correspondences. The initial matches with disparities and outliers are shown in Fig. 3.4. We apply the RANSAC paradigm to estimate quasi-fundamental matrix and remove the outliers. As shown in Fig. 3.4, all mismatches were rejected by the algorithm.

We calculated the average computation time in estimating the fundamental matrix under different models. Only linear algorithm was adopted and the minimal subsets for $\mathbf{F}$, $\mathbf{F}_q$, $\mathbf{F}_a$ are set as 8, 6, and 4 respectively. The program was implemented with Matlab R14 on Dell Inspiron 600 m laptop of Pentium(R) 1.8 GHz CPU. In first case, we select 200 correspondences and vary the outlier-to-inlier ratio from 0.1 to 1.0. In the second case, we set the outlier ratio at 0.8 and vary the feature number from 200 to 2000. The result is shown in Fig. 3.5. It is clear that the algorithm runs significantly faster with quasi-perspective model than perspective projection model, especially for larger data sets and higher outlier ratios.

3.5.3 Reconstruction Result

We reconstructed the 200 data points under different camera models according to the algorithm presented in Sect. 3.4. The reconstruction is defined up to a Euclidean rotation

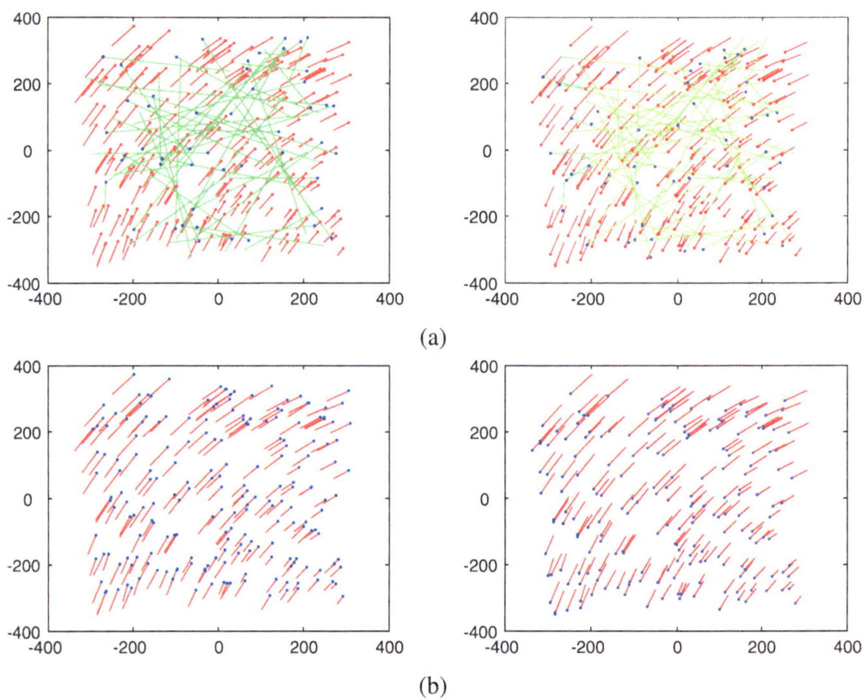

Fig. 3.4 Result of outlier removal. (**a**) Initial feature matches in two images with outliers; (**b**) Final detected correspondences after RANSAC algorithm

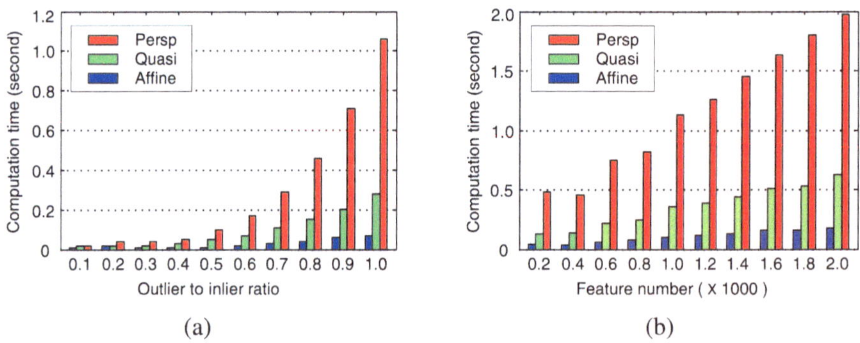

Fig. 3.5 Average computation time under different camera models. (**a**) The time with respect to different outlier-to-inlier ratios; (**b**) The time with respect to different feature point numbers

and translation with respect to the ground truth. We recover these parameters and register the result with the ground truth. The reconstruction error is defined as point-wise distance between the recovered structure and its ground truth. In order to obtain a statistically meaningful result, we vary the image noise level from 0 to 3 pixels in steps of 0.5, and

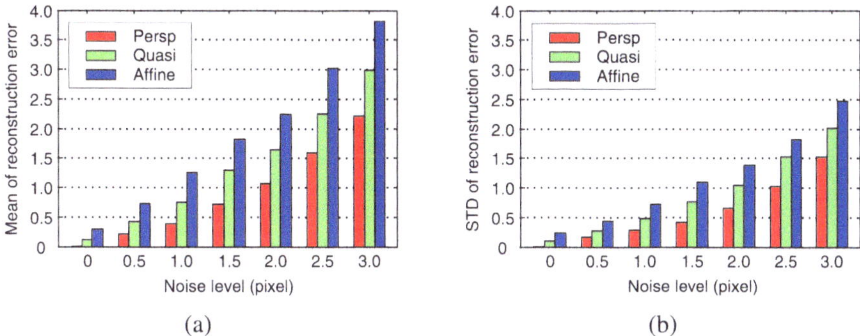

Fig. 3.6 Evaluation on 3D reconstruction accuracy by different models. (**a**) The mean of the reconstruction errors at different noise levels; (**b**) The corresponding standard deviation

take 100 independent tests at each noise level. The mean and standard deviation of the errors are shown in Fig. 3.6. Results validate that the reconstruction accuracy obtained using quasi-perspective model lies in between those obtained by perspective and affine projection models. The quasi-perspective model is more accurate than affine model.

3.6 Evaluations on Real Images

3.6.1 Test on Stone Dragon Images

We tested and compared different models on many real images and we will report two results here. The correspondences in the tests were established by the matching system based on SIFT and epipolar constraint [17], and the camera parameters were calibrated offline via the method in [24].

The stone dragon images were captured by Canon G3 camera in Chung Chi College of the Chinese University of Hong Kong. The image resolution is 1024×768 and 4261 reliable features were established by [17]. We recover the fundamental matrix via Gold Standard algorithm and reconstructed the 3D Euclidean structure according to the process in Sect. 3.4. Figure 3.7 shows the matched features, the reconstructed VRML model with texture mapping, and the corresponding wireframe model viewed from different viewpoints. The structure of the dragon is correctly recovered using the quasi-perspective model. The distributions of epipolar residual errors using the three models are compared in Fig. 3.8, the reprojection errors are listed in Table 3.3.

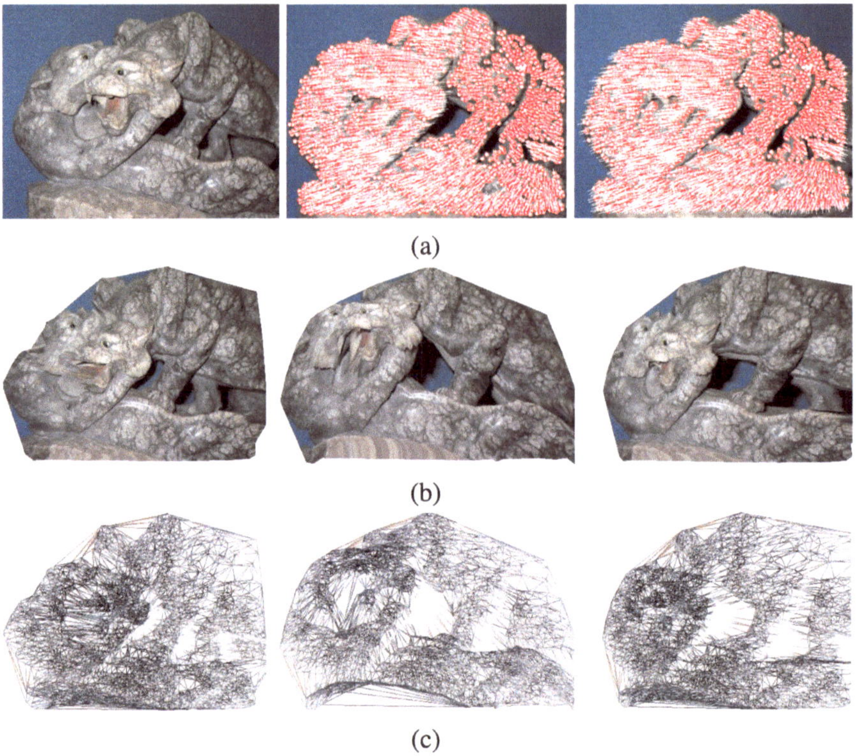

(a)

(b)

(c)

Fig. 3.7 Reconstruction result from stone dragon images. (**a**) Two images of the fountain base overlaid by tracked features with relative disparities; (**b**) Reconstructed VRML model of the scene shown from different viewpoints with texture mapping; (**c**) The corresponding triangulated wireframe of the VRML model

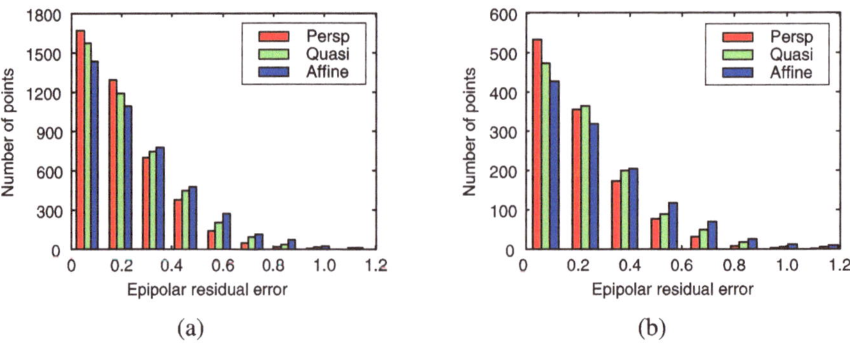

Fig. 3.8 Histogram distribution of the epipolar residual errors under different camera models. (**a**) Fountain base images; (**b**) Stone dragon images; (**c**) Medusa head images

Table 3.3 The reprojection error under different projection models

Model		Persp	Quasi	Affine
Error	Stone dragon	0.72	0.80	0.86
	Medusa head	0.97	1.04	1.13

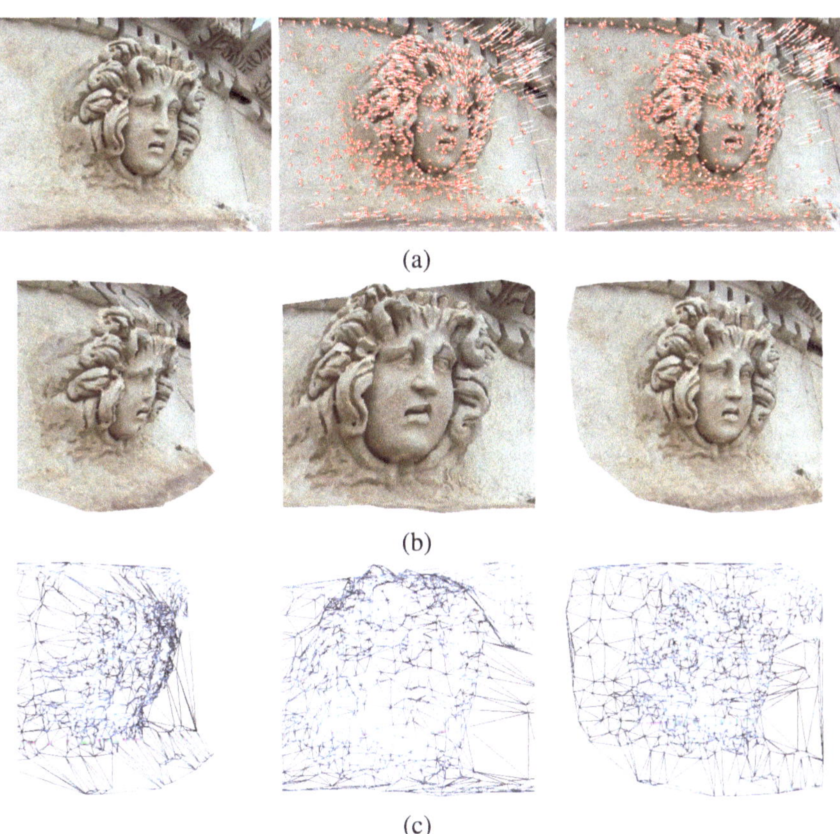

Fig. 3.9 Reconstruction result from Medusa head images. (**a**) Two images of the fountain base overlaid by tracked features with relative disparities; (**b**) Reconstructed VRML model of the scene shown from different viewpoints with texture mapping; (**c**) The corresponding triangulated wireframe of the VRML model

3.6.2 Test on Medusa Head Images

The Medusa head image sequence was downloaded from Dr. Pollefeys homepage which was taken in the ancient city of Sagalassos, Turkey. It was recorded by Sony TRV900 camcorder with a resolution of 720×576. We selected two adjacent frames and totally 1216 correspondences were established. The camera's intrinsic parameters were estimated by the method [19]. We recover the fundamental matrix and reconstruct its structure by the proposed algorithm. The reconstructed VRML model and the corresponding wireframe model are shown in Fig. 3.9. The result is realistic and visually correct despite of the unavailability of the ground truth. As a quantitative evaluation, we calculated the reprojection errors by the three projection models as tabulated in Table 3.3. The histogram distribution of the epipolar residual errors by different models are shown in Fig. 3.8. The result by quasi-perspective model is better than that by affine assumption.

3.7 Closure Remarks

3.7.1 Conclusion

In this chapter, we have investigated the one-view and two-view geometry of the quasi-perspective projection model and presented some special properties of the quasi fundamental matrix, plane induced homography, and 3D reconstruction under the proposed model. Both theoretical and experimental analysis show that quasi-perspective model is a good trade-off between the simplicity of affine and the accuracy of perspective projection. The result is consistent with our previous study on structure and motion factorization [18]. The proposed method is suitable to deal with images with small baseline.

3.7.2 Review Questions

1. *One-view geometry*. Show that the quasi-perspective projection matrix has 9 degrees of freedom. Under the projection, verify that parallel lines on the $O-XY$ coordinate plane or any other parallel planes are mapped to parallel lines in the image. Briefly describe why parallelism is not preserved in Z direction?
2. *Fundamental matrix*. Derive the expression of quasi-fundamental matrix and quasi-essential matrix. Why there are only 5 degrees of freedom? Given two quasi-perspective camera matrices $\mathbf{P}_q$ and $\mathbf{P}'_q$, show that the quasi-fundamental matrix between the two views can be expressed as $\mathbf{F}_q = [\mathbf{e}']_\times \mathbf{P}'_q \mathbf{P}_q^+$.
3. *Plane induced homography*. Explain the general form of plane induced homography under quasi-perspective assumption. Illustrate how you can recover homography from two pairs of correspondences with a given quasi-fundamental matrix.
4. *Structure reconstruction*. Give the form of a pair of canonical cameras under quasi-perspective projection. Show that two quasi-perspective reconstructions are defined up to a quasi-perspective transformation matrix.

References

1. Cheng, C.M., Lai, S.H.: A consensus sampling technique for fast and robust model fitting. Pattern Recogn. **42**(7), 1318–1329 (2009)
2. Dellaert, F., Seitz, S.M., Thorpe, C.E., Thrun, S.: Structure from motion without correspondence. In: Proc. of IEEE Conference on Computer Vision and Pattern Recognition, pp. 2557–2564 (2000)
3. Faugeras, O.: Stratification of 3-D vision: projective, affine, and metric representations. J. Opt. Soc. Am. A **12**, 465–484 (1995)
4. Fischler, M.A., Bolles, R.C.: Random sample consensus: A paradigm for model fitting with applications to image analysis and automated cartography. Commun. ACM **24**(6), 381–395 (1981)
5. Guilbert, N., Bartoli, A., Heyden, A.: Affine approximation for direct batch recovery of Euclidean structure and motion from sparse data. Int. J. Comput. Vis. **69**(3), 317–333 (2006)
6. Hartley, R.I.: In defense of the eight-point algorithm. IEEE Trans. Pattern Anal. Mach. Intell. **19**(6), 580–593 (1997)
7. Hartley, R.I., Zisserman, A.: Multiple View Geometry in Computer Vision, 2nd edn. Cambridge University Press, Cambridge (2004). ISBN: 0521540518
8. Hu, M., McMenemy, K., Ferguson, S., Dodds, G., Yuan, B.: Epipolar geometry estimation based on evolutionary agents. Pattern Recogn. **41**(2), 575–591 (2008)
9. Lehmann, S., Bradley, A.P., Clarkson, I.V.L., Williams, J., Kootsookos, P.J.: Correspondence-free determination of the affine fundamental matrix. IEEE Trans. Pattern Anal. Mach. Intell. **29**(1), 82–97 (2007)
10. Mendonça, P.R.S., Cipolla, R.: Analysis and computation of an affine trifocal tensor. In: Proc. of British Machine Vision Conference, pp. 125–133 (1998)
11. Mundy, J.L., Zisserman, A.: Geometric Invariance in Computer Vision. MIT Press, Cambridge (1992)
12. Oliensis, J., Hartley, R.: Iterative extensions of the Sturm/Triggs algorithm: Convergence and nonconvergence. IEEE Trans. Pattern Anal. Mach. Intell. **29**(12), 2217–2233 (2007)
13. Torr, P.H.S., Zisserman, A., Maybank, S.J.: Robust detection of degenerate configurations while estimating the fundamental matrix. Comput. Vis. Image Underst. **71**(3), 312–333 (1998)
14. Poelman, C., Kanade, T.: A paraperspective factorization method for shape and motion recovery. IEEE Trans. Pattern Anal. Mach. Intell. **19**(3), 206–218 (1997)
15. Shapiro, L.S., Zisserman, A., Brady, M.: 3D motion recovery via affine epipolar geometry. Int. J. Comput. Vis. **16**(2), 147–182 (1995)
16. Shimshoni, I., Basri, R., Rivlin, E.: A geometric interpretation of weak-perspective motion. IEEE Trans. Pattern Anal. Mach. Intell. **21**(3), 252–257 (1999)
17. Wang, G.: A hybrid system for feature matching based on sift and epipolar constraints. Tech. Rep. Department of ECE, University of Windsor (2006)
18. Wang, G., Wu, J.: Quasi-perspective projection with applications to 3D factorization from uncalibrated image sequences. In: Proc. of IEEE Conference on Computer Vision and Pattern Recognition, pp. 1–8 (2008)
19. Wang, G., Wu, J.: Perspective 3D Euclidean reconstruction with varying camera parameters. IEEE Trans. Circuits Syst. Video Technol. **19**(12), 1793–1803 (2009)
20. Wang, G., Wu, J.: The quasi-perspective model: Geometric properties and 3D reconstruction. Pattern Recogn. **43**(5), 1932–1942 (2010)
21. Wang, G., Wu, J.: Quasi-perspective projection model: Theory and application to structure and motion factorization from uncalibrated image sequences. Int. J. Comput. Vis. **87**(3), 213–234 (2010)
22. Weng, J., Huang, T., Ahuja, N.: Motion and structure from two perspective views: Algorithms. IEEE Trans. Pattern Anal. Mach. Intell. **11**(5), 451–476 (1997)
23. Wolf, L., Shashua, A.: Affine 3-D reconstruction from two projective images of independently translating planes. In: Proc. of International Conference on Computer Vision, pp. 238–244 (2001)

3

24. Zhang, Z.: A flexible new technique for camera calibration. IEEE Trans. Pattern Anal. Mach. Intell. **22**(11), 1330–1334 (2000)
25. Zhang, Z., Anandan, P., Shum, H.Y.: What can be determined from a full and a weak perspective image? In: Proc. of International Conference on Computer Vision, pp. 680–687 (1999)
26. Zhang, Z., Kanade, T.: Determining the epipolar geometry and its uncertainty: A review. Int. J. Comput. Vis. **27**(2), 161–195 (1998)
27. Zhang, Z., Xu, G.: A general expression of the fundamental matrix for both projective and affine cameras. In: Proc. of International Joint Conference on Artificial Intelligence, pp. 1502–1507 (1997)

Introduction to Structure and Motion Factorization 4

Abstract The chapter reviews some popular algorithms for structure and motion factorization. We first introduce the problem of structure and motion recovery. Then, present the main idea of the following factorization algorithms for different kinds of scenarios under different projection models. (i) Structure and motion factorization of rigid objects under affine assumption and its extension to perspective camera model; (ii) Nonrigid factorization under both affine and perspective projection model; (iii) Structure factorization of multiple and articulated objects.

If I have been able to see further, it was only because I stood on the shoulders of giants.

Sir Isaac Newton (1643–1727)

4.1 Introduction

The problem of structure and motion recovery from image sequences is an important theme in computer vision. Great progress have been made for different applications during the last two decades [17]. Among these methods, factorization based approach, for its robust behavior and accuracy, is widely studied since it deals uniformly with the data sets of all images [28, 29, 35, 39]. The factorization algorithm was first proposed by Tomasi and Kanade [35] in the early 90's. The main idea of this algorithm is to factorize the tracking matrix into motion and structure matrices simultaneously by singular value decomposition (SVD) with low-rank approximation. The algorithm assumes an orthographic projection model. It was extended to weak perspective and paraperspective projection by Poelman and Kanade [28]. The orthographic, weak perspective, and paraperspective projections can be generalized as affine camera models.

More generally, Christy and Horaud [8] extended the above methods to a perspective camera model by incrementally performing the factorization under affine assumption. The method is an affine approximation to full perspective projection. Triggs and Sturm [33, 39]

G. Wang, Q.M.J. Wu, *Guide to Three Dimensional Structure and Motion Factorization*, Advances in Pattern Recognition,
DOI 10.1007/978-0-85729-046-5_4,

proposed a full projective reconstruction method via rank-4 factorization of a scaled tracking matrix with projective depths recovered from pairwise epipolar geometry. The method was further studied in [15, 18, 24], where different iterative schemes were proposed to recover the projective depths through minimization of reprojection errors. Recently, Oliensis and Hartley [26] provided a complete theoretical convergence analysis for the iterative extensions. Unfortunately, no iteration has been shown to converge sensibly, and they proposed a simple extension, called CIESTA, to give a reliable initialization to other algorithms.

The above methods work only for rigid objects and static scenes. Whereas in real world, many scenarios are nonrigid or dynamic such as articulated motion, human faces carrying different expressions, lip movements, hand gesture, and moving vehicles etc. In order to deal with such situations, many extensions stemming from the factorization algorithm were proposed to relax the rigidity constraint. Costeira and Kanade [9] first discussed how to recover the motion and shape of several independent moving objects via factorization using orthographic projection. Bascle and Blake [3] proposed a method for factorizing facial expressions and poses based on a set of preselected basis images. Recently, Li *et al.* [21] proposed to segment multiple rigid-body motions from point correspondences via subspace separation. Yan and Pollefeys [46, 47] proposed a factorization-based approach to recover the structure and kinematic chain of articulated objects. Zelnik-Manor and Irani [49, 50] analyzed the problem of multi-sequence factorization of multiple objects by both temporal synchronization of sequences and spacial matching across sequences. Del Bue and Agapito [11] proposed a scheme for nonrigid stereo factorization.

In the pioneer work by Bregler *et al.* [6], it is demonstrated that the 3D shape of a nonrigid object can be expressed as a weighted linear combination of a set of shape bases. Then the shape bases and camera motions are factorized simultaneously for all time instants under the rank constraint of the tracking matrix. Following this idea, the method was extensively investigated and developed by many researchers, such as Brand [4, 5], Del Bue *et al.* [12, 13], Torresani *et al.* [36, 37], and Xiao *et al.* [43, 45]. Recently, Rabaud and Belongie [31] relaxed the Bregler's assumption [6] by assuming that only small neighborhoods of shapes are well-modeled by a linear subspace, and proposed a novel approach to solve the problem by adopting a manifold-learning framework.

Most nonrigid factorization methods are based on affine camera model due to its simplicity. It was extended to perspective projection in [45] by iteratively recovering the projective depths. The perspective factorization is more complicated and does not guarantee its convergence to the correct depths, especially for nonrigid scenarios [17]. Vidal and Abretske [40] proposed that the constraints among multiple views of a nonrigid shape consisting of k shape bases can be reduced to multilinear constraints. They presented a closed form solution to the reconstruction of a nonrigid shape consisting of two shape bases. Hartley and Vidal [16] proposed a closed form solution to the nonrigid shape and motion with calibrated cameras or fixed intrinsic parameters. Since the factorization is only defined up to a nonsingular transformation matrix, many researchers adopt the metric constraints to recover the matrix and upgrade the factorization to the Euclidean space [4, 6, 13, 37]. However, the rotation constraint may cause ambiguity in the combination of shape bases. Xiao *et al.* [43] proposed a basis constraint to solve the ambiguity and provide a closed-form solution to the problem.

Above factorization methods usually assume error-free tracking data. The performance will degrade in present of outliers or large errors. More recently, the problem of robust factorization has been concerned and some powerful methods have been proposed to deal with noisy and erroneous data [48]. Anandan and Irani [2] proposed a covariance-weighted factorization to factor noisy correspondences with high degree of directional uncertainty into structure and motion. Gruber and Weiss [14] formulated the problem as one of factor analysis and derived an EM algorithm to incorporate prior knowledge and enhance the robustness to missing data and uncertainties. Zelnik-Manor *et al.* [51] defined a new type of motion consistency based on temporal consistency, and applied it to multi-body factorization with directional uncertainty. Zaharescu and Horaud [48] introduced a Gaussian/uniform mixture model and associated EM algorithm that is resilient to outliers. Buchanan and Fitzgibbon [7] presented a comprehensive comparison on many factorization algorithms. Their study strongly support the second order nonlinear optimization strategies. Some more studies can be found in [1, 25].

The remaining part of this chapter is organized as follows. We first introduce the problem of structure and motion recovery in Sect. 4.2. In Sect. 4.3, we present the general algorithms for structure and motion factorization of rigid objects under both affine and perspective camera models. Then we introduce the factorization algorithm for nonrigid objects in Sect. 4.4. Finally, methods for multi-body and articulated object factorization are discussed in Sect. 4.5.

4.2 Problem Definition

As discussed in Chap. 2, a general imaging process under perspective projection is formulated as

$$\lambda_{ij}\mathbf{x}_{ij} = \mathbf{P}_i\mathbf{X}_j = \mathbf{K}_i[\mathbf{R}_i, \mathbf{T}_i]\mathbf{X}_j \tag{4.1}$$

where λ_{ij} is a nonzero depth scale, $\mathbf{P}_i$ is a 3×4 projection matrix of ith camera. When the object is far away from the camera, we may assume a simplified affine camera model to approximate perspective projection.

$$\bar{\mathbf{x}}_{ij} = \mathbf{A}_i\bar{\mathbf{X}}_j + \bar{\mathbf{T}}_i \tag{4.2}$$

where the affine matrix $\mathbf{A}_i$ is a 2×3 matrix; the image point $\bar{\mathbf{x}}_{ij}$ and space point $\bar{\mathbf{X}}_j$ are denoted in nonhomogeneous form. Under affine projection, the mapping from space to the image becomes linear since the unknown depth scalar λ_{ij} in (4.1) is removed. If we adopt relative coordinates and register all image points to the centroid, the translation term $\bar{\mathbf{T}}_i$ can be eliminated. Thus, the affine projection (4.2) is simplified to

$$\bar{\mathbf{x}}_{ij} = \mathbf{A}_i\bar{\mathbf{X}}_j \tag{4.3}$$

Given an image sequence of m frames. Suppose we have a set of n feature points tracked across the sequence with coordinates $\{\bar{\mathbf{x}}_{ij} | i = 1, \ldots, m, j = 1, \ldots, n\}$, where the image

4

point $\bar{\mathbf{x}}_{ij} = [u_{ij}, v_{ij}]^T$ is expressed in non-homogeneous form. For all these features, we can arrange them into a compact single matrix form as:

$$\mathbf{W} = \texttt{frames}\downarrow \overbrace{\begin{bmatrix} \bar{\mathbf{x}}_{11} & \cdots & \bar{\mathbf{x}}_{1n} \\ \vdots & \ddots & \vdots \\ \bar{\mathbf{x}}_{m1} & \cdots & \bar{\mathbf{x}}_{mn} \end{bmatrix}}^{\texttt{points}}_{2m\times n} = \begin{bmatrix} \begin{bmatrix} u_{11} \\ v_{11} \end{bmatrix} & \cdots & \begin{bmatrix} u_{1n} \\ v_{1n} \end{bmatrix} \\ \vdots & \ddots & \vdots \\ \begin{bmatrix} u_{m1} \\ v_{m1} \end{bmatrix} & \cdots & \begin{bmatrix} u_{mn} \\ v_{mn} \end{bmatrix} \end{bmatrix} \tag{4.4}$$

where $\mathbf{W}$ is called the tracking or measurement matrix, a $2m \times n$ matrix composed of all tracking information across the sequence. Under perspective projection (4.1), if we include the depth scale, the tracking data can be denoted in homogeneous form as follows.

$$\dot{\mathbf{W}} = \begin{bmatrix} \lambda_{11}\mathbf{x}_{11} & \cdots & \lambda_{1n}\mathbf{x}_{1n} \\ \vdots & \ddots & \vdots \\ \lambda_{m1}\mathbf{x}_{m1} & \cdots & \lambda_{mn}\mathbf{x}_{mn} \end{bmatrix}_{3m\times n} = \begin{bmatrix} \lambda_{11}\begin{bmatrix} u_{11} \\ v_{11} \\ 1 \end{bmatrix} & \cdots & \lambda_{1n}\begin{bmatrix} u_{1n} \\ v_{1n} \\ 1 \end{bmatrix} \\ \vdots & \ddots & \vdots \\ \lambda_{m1}\begin{bmatrix} u_{m1} \\ v_{m1} \\ 1 \end{bmatrix} & \cdots & \lambda_{mn}\begin{bmatrix} u_{mn} \\ v_{mn} \\ 1 \end{bmatrix} \end{bmatrix} \tag{4.5}$$

We call $\dot{\mathbf{W}}$ the projective-depth-scaled or weighted tracking matrix. The depth scale λ_{ij} is normally unknown.

Figure 4.1 gives an example of the feature tracking result for a dinosaur sequence. The sequence contains 36 frames which are taken evenly around a turn table. The images and tracking data were downloaded from the Visual Geometry Group of Oxford University. Figure 4.1 shows 4 consecutive images with 115 tracked points. Feature tracking has been an active research topic since the beginning of computer vision. Many effective

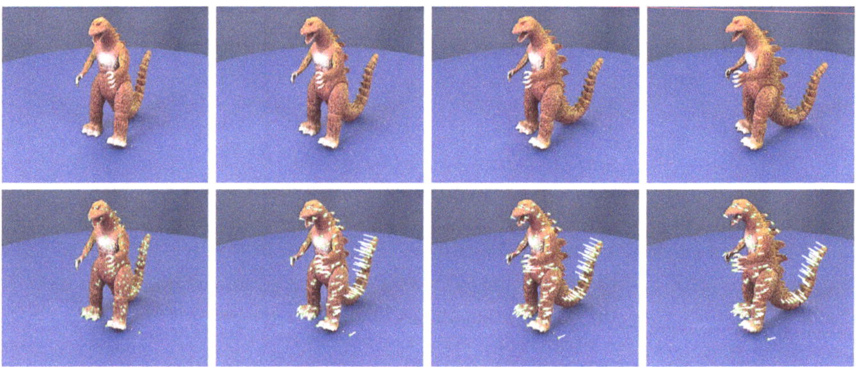

Fig. 4.1 Four consecutive frames from the dinosaur sequence. Where 115 tracked features are overlaid on the images in the *second row*, the *white lines* denote the relative disparities between frames. Courtesy of Andrew Fitzgibbon and Andrew Zisserman

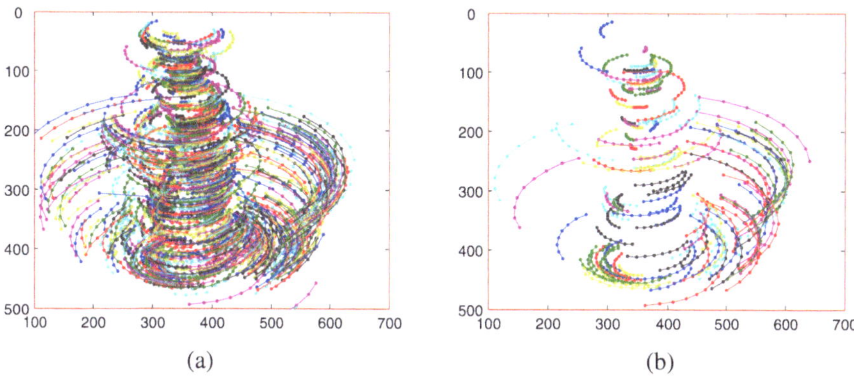

Fig. 4.2 The track of features in dinosaur sequence. (**a**) The features being tracked in more than four consecutive frames; (**b**) The features being tracked in more than eight consecutive frames. The consecutive tracked features are connected by color lines

Table 4.1 Classification of structure and motion factorization of rigid and nonrigid objects. 'Persp.' stands for perspective projection; 'Quasi' stands for quasi-perspective projection

Classification		Tracking	Motion	Shape	Upgrading matrix
	Affine	$\mathbf{W} \in \mathbb{R}^{2m \times n}$	$\mathbf{M} \in \mathbb{R}^{2m \times 3}$	$\bar{\mathbf{S}} \in \mathbb{R}^{3 \times n}$	$\mathbf{H} \in \mathbb{R}^{3 \times 3}$
Rigid	Persp.	$\dot{\mathbf{W}} \in \mathbb{R}^{3m \times n}$	$\mathbf{M} \in \mathbb{R}^{3m \times 4}$	$\mathbf{S} \in \mathbb{R}^{4 \times n}$	$\mathbf{H} \in \mathbb{R}^{4 \times 4}$
	Quasi	$\mathbf{W} \in \mathbb{R}^{3m \times n}$	$\mathbf{M} \in \mathbb{R}^{3m \times 4}$	$\mathbf{S} \in \mathbb{R}^{4 \times n}$	$\mathbf{H} \in \mathbb{R}^{4 \times 4}$
	Affine	$\mathbf{W} \in \mathbb{R}^{2m \times n}$	$\mathbf{M} \in \mathbb{R}^{2m \times 3k}$	$\bar{\mathbf{B}} \in \mathbb{R}^{3k \times n}$	$\mathbf{H} \in \mathbb{R}^{3k \times 3k}$
Nonrigid	Persp.	$\dot{\mathbf{W}} \in \mathbb{R}^{3m \times n}$	$\mathbf{M} \in \mathbb{R}^{3m \times (3k+1)}$	$\mathbf{B} \in \mathbb{R}^{(3k+1) \times n}$	$\mathbf{H} \in \mathbb{R}^{(3k+1) \times (3k+1)}$
	Quasi	$\mathbf{W} \in \mathbb{R}^{3m \times n}$	$\mathbf{M} \in \mathbb{R}^{3m \times (3k+1)}$	$\mathbf{B} \in \mathbb{R}^{(3k+1) \times n}$	$\mathbf{H} \in \mathbb{R}^{(3k+1) \times (3k+1)}$

methods have been proposed, such as the Kanade-Lucas-Tomasi (KLT) Feature Tracker [23, 32], Stereo vision matching via epipolar geometry [52], Scale-invariant feature transform (SIFT) [22], and many more.

In practice, however, some features may not be tracked in certain frames due to self occlusion or tracking failure, this will leave some holes (missing data) in the tracking matrix. Besides, the data may be corrupted by large image noise or outliers. Figure 4.2 shows the track of all tracked features in the dinosaur sequence, from which we can see that many features have been lost during the tracking process. In this book, we usually assume the ideal case without missing data and outliers.

The problem of structure from motion is defined as: given the tracked features across a sequence, we want to recover the 3D Euclidean structure $\mathbf{S}_{ij} = \{\mathbf{X}_{ij}\}$ for $i = 1, \dots, m$, $j = 1, \dots, n$, and motion parameters $\{\mathbf{R}_i, \mathbf{T}_i\}$ of the object corresponding to each frame. The factorization based algorithm has been proved to be an effective method to deal with this problem. According to the property of the object and the camera model employed, the algorithm can generally be classified into the following categories as shown in Table 4.1.

1. Structure and motion factorization of rigid objects under affine assumption;

4

2. Factorization of rigid objects under perspective projection model;
3. Factorization of rigid objects under quasi-perspective projection model;
4. Factorization of nonrigid objects under affine projection model;
5. Factorization of nonrigid objects under perspective projection model;
6. Factorization of nonrigid objects under quasi-perspective projection model.

In Table 4.1, the meaning of some symbols, such as **M**, **S**, **B**, and **H** will be defined in the following sections. The quasi-perspective factorization algorithm [42] will be discussed in Chap. 9 of the book.

4.3 Structure and Motion Factorization of Rigid Objects

In this section, we will introduce the structure and motion factorization algorithm for rigid objects under affine and perspective camera model.

4.3.1 Rigid Factorization Under Orthographic Projection

Suppose the image points in each frame are registered to the corresponding centroid. Under orthographic projection, the imaged points in the ith frame can be formulated as follows.

$$[\bar{\mathbf{x}}_{i1}, \bar{\mathbf{x}}_{i2}, \ldots, \bar{\mathbf{x}}_{in}] = \mathbf{R}_{Ai}[\bar{\mathbf{X}}_1, \bar{\mathbf{X}}_2, \ldots, \bar{\mathbf{X}}_n], \quad \forall i = 1, \ldots, m \tag{4.6}$$

If we stack the equations (4.6) for all frames together, we can obtain

$$\underbrace{\begin{bmatrix} \bar{\mathbf{x}}_{11} & \cdots & \bar{\mathbf{x}}_{1n} \\ \vdots & \ddots & \vdots \\ \bar{\mathbf{x}}_{m1} & \cdots & \bar{\mathbf{x}}_{mn} \end{bmatrix}}_{\mathbf{W}_{2m\times n}} = \underbrace{\begin{bmatrix} \mathbf{R}_{A1} \\ \vdots \\ \mathbf{R}_{Am} \end{bmatrix}}_{\mathbf{M}_{2m\times 3}} \underbrace{[\bar{\mathbf{X}}_1, \ldots, \bar{\mathbf{X}}_n]}_{\bar{\mathbf{S}}_{3\times n}} \tag{4.7}$$

The above equation is called general factorization expression under orthographic projection, which can be written concisely as $\mathbf{W} = \mathbf{M}\bar{\mathbf{S}}$. Suppose the tracking data across the sequence of m frames are available, i.e. the tracking matrix $\mathbf{W}$ is given, our purpose is to recover the motion matrix $\mathbf{M}$ and shape matrix $\bar{\mathbf{S}}$.

4.3.1.1 SVD Decomposition with Rank Constraint

The tracking matrix is a $2m \times n$ matrix with highly rank-deficiency. From the right side of (4.7) we can easily find that the rank of the tracking matrix is at most 3 for noise free data since both $\mathbf{M}$ and $\bar{\mathbf{S}}$ are at most of rank 3. However, when the data is corrupted by image noise, $\mathbf{W}$ will not be exactly rank-3. Here we use the concept of SVD decomposition to obtain the rank-3 approximation and factorize the tracking matrix into the motion and shape matrices.

Without loss of generosity, let us assume $2m \geq n$ and perform SVD decomposition on the tracking matrix.

$$\mathbf{W} = \mathbf{U}_{2m\times n}\Sigma_{n\times n}\mathbf{V}_{n\times n}^T \tag{4.8}$$

where $\Sigma = \text{diag}(\sigma_1, \sigma_2, \ldots, \sigma_n)$ is a diagonal matrix with all diagonal entries composed by the singular values of $\mathbf{W}$ and arranged in descending order as $\sigma_1 \geq \sigma_2 \geq \cdots \geq \sigma_n \geq 0$. $\mathbf{U}$ and $\mathbf{V}$ are $2m \times n$ and $n \times n$ orthogonal matrices respectively, thus $\mathbf{U}^T\mathbf{U} = \mathbf{V}^T\mathbf{V} = \mathbf{I}_n$ with $\mathbf{I}_n$ a $n \times n$ identical matrix. It is noted that the assumption $2m \geq n$ is not crucial, we can obtain a similar decomposition when $2m < n$ by simply taking a transpose of the tracking matrix.

In ideal case, $\mathbf{W}$ is of rank 3, which is equivalent to $\sigma_4 = \sigma_5 = \cdots = \sigma_n = 0$. When the data is contaminated by noise, the rank of $\mathbf{W}$ is definitely greater than 3. Actually, the rank may also be greater than 3 even for noise free data since the affine camera model is just an approximation of the real imaging process. We will now seek a rank-3 matrix $\mathbf{W}'$ that can best approximate the tracking matrix. Let us partition the matrices $\mathbf{U}$, Σ, and $\mathbf{V}$ as follows.

$$\begin{aligned}
\mathbf{U} &= [\mathbf{U}'_{2m\times 3} | \mathbf{U}''_{2m\times(n-3)}] \\
\Sigma &= \begin{bmatrix} \Sigma'_{3\times 3} & \mathbf{0} \\ \mathbf{0} & \Sigma''_{(n-3)\times(n-3)} \end{bmatrix} \\
\mathbf{V} &= [\mathbf{V}'_{n\times 3} | \mathbf{V}''_{n\times(n-3)}]
\end{aligned} \tag{4.9}$$

then the SVD decomposition (4.8) can be written as

$$\mathbf{W} = \underbrace{\mathbf{U}'\Sigma'\mathbf{V}'^T}_{\mathbf{W}'} + \underbrace{\mathbf{U}''\Sigma''\mathbf{V}''^T}_{\mathbf{W}''} \tag{4.10}$$

where $\Sigma' = \text{diag}(\sigma_1, \sigma_2, \sigma_3)$ contains the first three greatest singular values of the tracking matrix, $\mathbf{U}'$ is the first three columns of $\mathbf{U}$, $\mathbf{V}'^T$ is the first three rows of $\mathbf{V}^T$. It is easy to prove that $\mathbf{W}' = \mathbf{U}'\Sigma'\mathbf{V}'^T$ is the best rank-3 approximation of $\mathbf{W}$ in the Frobenius norm. Now let us define

$$\tilde{\mathbf{M}} = \mathbf{U}'\Sigma'^{\frac{1}{2}} \tag{4.11}$$

$$\tilde{\mathbf{S}} = \Sigma'^{\frac{1}{2}}\mathbf{V}'^T \tag{4.12}$$

where $\tilde{\mathbf{M}}$ is a $2m \times 3$ matrix and $\tilde{\mathbf{S}}$ is a $3 \times n$ matrix. Then we have $\mathbf{W}' = \tilde{\mathbf{M}}\tilde{\mathbf{S}}$, a similar form of the factorization expression (4.7). In fact, $\tilde{\mathbf{M}}$ and $\tilde{\mathbf{S}}$ are one set of the maximum likelihood affine reconstruction of the tracking matrix $\mathbf{W}$. However, the decomposition is not unique since it is only defined up to a nonsingular linear transformation matrix $\mathbf{H} \in \mathbb{R}^{3\times 3}$ as $\tilde{\mathbf{M}}\tilde{\mathbf{S}} = (\tilde{\mathbf{M}}\mathbf{H})(\mathbf{H}^{-1}\tilde{\mathbf{S}})$. If we can find a transformation matrix $\mathbf{H}$ that can

4

make

$$\mathbf{M} = \tilde{\mathbf{M}}\mathbf{H} \tag{4.13}$$

exactly corresponds to a Euclidean motion matrix as in (4.7), then the structure $\bar{\mathbf{S}} = \mathbf{H}^{-1}\tilde{\mathbf{S}}$ will be upgraded from affine to Euclidean space. We call the transformation $\mathbf{H}$ upgrading matrix, which can be recovered by enforcing orthogonality of the rotation matrix, which is also named as metric constraint.

4.3.1.2
Euclidean Stratification and Reconstruction

Suppose there is an upgrading matrix $\mathbf{H}$ that upgrades the matrix $\tilde{\mathbf{M}}$ in (4.11) to the Euclidean motion matrix in (4.7), then the motion matrix corresponding to frame i can be written as

$$\begin{bmatrix} \mathbf{r}_{1i}^T \\ \mathbf{r}_{2i}^T \end{bmatrix} = \mathbf{M}_i = \tilde{\mathbf{M}}_i\mathbf{H} = \begin{bmatrix} \mathbf{m}_{1i}^T \\ \mathbf{m}_{2i}^T \end{bmatrix} \mathbf{H} \tag{4.14}$$

which leads to

$$\begin{bmatrix} \mathbf{m}_{1i}^T \\ \mathbf{m}_{2i}^T \end{bmatrix} \mathbf{H}\mathbf{H}^T \left[\mathbf{m}_{1i} | \mathbf{m}_{2i}\right] = \begin{bmatrix} \mathbf{r}_{1i}^T \\ \mathbf{r}_{2i}^T \end{bmatrix} \left[\mathbf{r}_{1i} | \mathbf{r}_{2i}\right] = \begin{bmatrix} 1 & 0 \\ 0 & 1 \end{bmatrix} \tag{4.15}$$

Let us define $\mathbf{Q} = \mathbf{H}\mathbf{H}^T$ which is a 3×3 symmetric matrix with 6 unknowns. Then we can obtain the following constraints.

$$\begin{cases} \mathbf{m}_{1i}^T\mathbf{Q}\mathbf{m}_{1i} = 1 \\ \mathbf{m}_{2i}^T\mathbf{Q}\mathbf{m}_{2i} = 1 \\ \mathbf{m}_{1i}^T\mathbf{Q}\mathbf{m}_{2i} = \mathbf{m}_{2i}^T\mathbf{Q}\mathbf{m}_{1i} = 0 \end{cases} \tag{4.16}$$

The constraints (4.16) are called metric or rotation constraints, which yield a set of over-constrained equations for all frames $i = 1, \ldots, m$. Thus $\mathbf{Q}$ can be calculated linearly via least squares and the upgrading matrix $\mathbf{H}$ is then extracted from $\mathbf{Q}$ using Cholesky decomposition [17]. Finally, the correct metric motion and structure matrices are obtained by applying the upgrading matrix as $\mathbf{M} = \tilde{\mathbf{M}}\mathbf{H}$, $\bar{\mathbf{S}} = \mathbf{H}^{-1}\tilde{\mathbf{S}}$, and the rotation matrices corresponding to each frame are then extracted from $\mathbf{M}$.

It is noted that the above solution is only defined up to an arbitrary rotation matrix since the choice of world coordinate system is free. In practice, we can simply choose the first frame as a reference and register all other frames to it, which is equivalent to set $\mathbf{R}_1 = \mathbf{I}_3$.

The implementation details of the above factorization algorithm are summarized as follows.

1. Register all image points in each frame to its centroid and construct the tracking matrix;
2. Perform rank-3 SVD factorization on the tracking matrix to obtain a solution of $\hat{\mathbf{M}}$ and $\hat{\mathbf{S}}$ from (4.11) and (4.12);
3. Compute the upgrading matrix $\mathbf{H}$ from (4.16);
4. Recover the Euclidean motion matrix $\mathbf{M} = \hat{\mathbf{M}}\mathbf{H}$ and shape matrix $\mathbf{S} = \mathbf{H}^{-1}\hat{\mathbf{S}}$;
5. Retrieve the rotation matrix of each frame from $\mathbf{M}$.

Remark 4.1 The above algorithm assumes orthographic projection model, which is a simple approximation when the thickness and depth variation of the object between frames are very small compared to the distance from the object to the cameras. The method has later been extended to more general affine camera models, such as weak-perspective [19] and para-perspective projection model [28]. For more general affine cameras, Quan [29] presented a study on the metric constraints and camera self-calibration.

Remark 4.2 Almost all factorization algorithms are limited in handling the tracking data of point features. Alternatively, Quan and Kanade [30] proposed an analogous factorization algorithm for line features under affine assumption. The algorithm decomposes the whole structure and motion into three substructures which can be solved linearly via factorization of appropriate measurement matrix. The line-based factorization requires at least seven lines in three views. Whereas in point-based affine factorization, we only need a minimum of four points in three frames.

4.3.2 Rigid Factorization Under Perspective Projection

Many previous studies on rigid factorization adopt affine camera model due to its simplicity. However, the assumption is only valid when the objects have small depth variation and are far away from the cameras. Otherwise, the algorithm may fail or give poor results. More generally, Christy and Horaud [8] extended the above methods to a perspective camera model by incrementally performing the factorization under affine assumption. The method is an affine approximation to full perspective projection. Sturm [33] and Triggs and Sturm [39] proposed a full projective reconstruction method via rank-4 factorization of a scaled tracking matrix with projective depths recovered from pairwise epipolar geometry. The method was further studied in [15, 18, 24], where different iterative schemes were proposed to recover the projective depths by minimizing image reprojection errors.

Under perspective projection (4.1), all imaged points in the ith frame are formulated as

$$[\mathbf{x}_{i1}, \mathbf{x}_{i2}, \dots, \mathbf{x}_{in}] = \mathbf{P}_i[\mathbf{X}_1, \mathbf{X}_2, \dots, \mathbf{X}_n], \quad \forall i = 1, \dots, m \tag{4.17}$$

4

Thus we can obtain the general perspective factorization expression by gathering (4.17) for all frames as follows.

$$\underbrace{\begin{bmatrix} \lambda_{11}\mathbf{x}_{11} & \cdots & \lambda_{1n}\mathbf{x}_{1n} \\ \vdots & \ddots & \vdots \\ \lambda_{m1}\mathbf{x}_{m1} & \cdots & \lambda_{mn}\mathbf{x}_{mn} \end{bmatrix}}_{\dot{\mathbf{W}}_{3m\times n}} = \underbrace{\begin{bmatrix} \mathbf{P}_1 \\ \vdots \\ \mathbf{P}_m \end{bmatrix}}_{\mathbf{M}_{3m\times 4}} \underbrace{[\mathbf{X}_1, \ldots, \mathbf{X}_n]}_{\mathbf{S}_{4\times n}} \tag{4.18}$$

Compared with affine factorization (4.7), the main differences lie in the dimension and entries of the tracking, motion, and shape matrices. If a set of correct projective scales λ_{ij} are available, then the rank of the weighted tracking matrix is at most 4 as the rank of either the motion matrix $\mathbf{M}$ or the shape matrix $\mathbf{S}$ is not greater than 4. For noise contaminated data, $rank(\dot{\mathbf{W}}) > 4$, and we can adopt a similar SVD decomposition process as in (4.9) to obtain a best rank-4 approximation of the scale-weighted tracking matrix and factorize it into a $3m \times 4$ motion matrix $\tilde{\mathbf{M}}$ and a $4 \times n$ shape matrix $\tilde{\mathbf{S}}$. Obviously, such factorization corresponds to a projective reconstruction, which is defined up to a 4×4 transformation matrix $\mathbf{H}$ to upgrade the solution from projective to Euclidean space.

Through above analysis, we can see that there are essentially two complications in perspective factorization. One is the computation of the projective depths, the other is the recovery of the upgrading matrix. We will give a further discussion on these problems in next chapter.

4.4 Structure and Motion Factorization of Nonrigid Objects

We assumed rigid objects and static scenes in last section. While in real world, many objects do not have fixed structure, such as human faces with different expressions, torsos, and animals bodies, etc. In this section, we will extend the factorization algorithm to such nonrigid and deformable objects. Figure 4.3 shows the deformable structure of a jellyfish at different time instance.

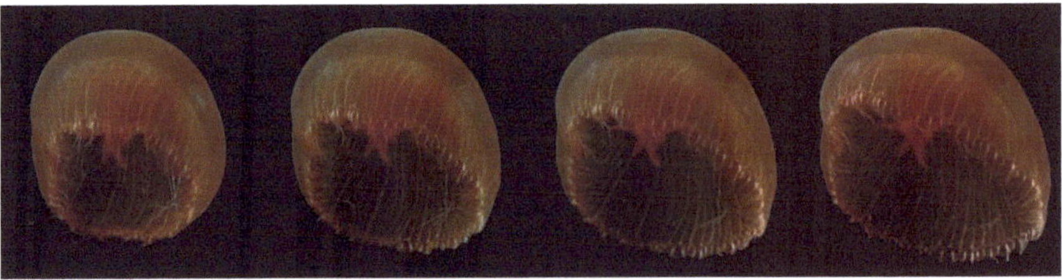

Fig. 4.3 The structure of a jellyfish with different deformations. Courtesy of BBC Planet Earth TV series

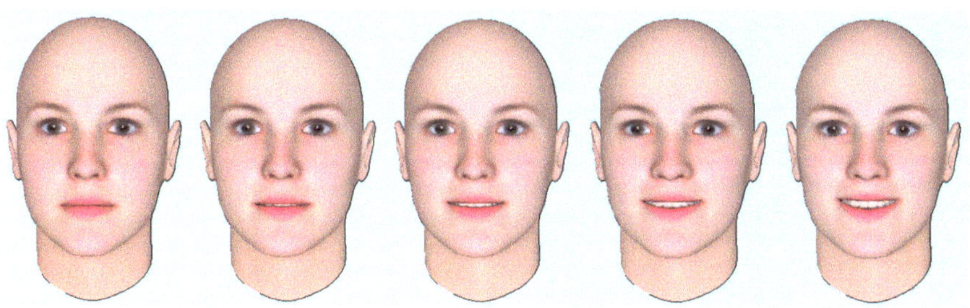

Fig. 4.4 Five female face models carrying expressions from neutral to smiling. We may take any two models as shape bases, then the other models can be derived as weighted linear combinations of the two bases. The models are generated by FaceGen Modeller http://www.facegen.com/index.htm

4.4.1 Bregler's Deformation Model

For nonrigid object, if all surface features deform randomly at any time instance, there is currently no suitable method to recover its structure from images. Here we restrict our study to a specific kind of deformation objects following the idea first proposed by Bregler *et al.* [6], where the 3D structure of nonrigid object is approximated by a weighted combination of a set of shape bases. Figure 4.4 shows a very simple example of face models from neutral to smiling with only mouth movements. The deformation structure can be modeled from only two shape bases. If more face expressions, such as joy, sadness, surprise, fear, etc., are involved, then more shape bases are needed to model the structure.

Suppose the deformation structure $\bar{\mathbf{S}}_i \in \mathbb{R}^{3\times n}$ is expressed as a weighted combination of k principal modes of deformation $\mathbf{B}_l \in \mathbb{R}^{3\times n}, l = 1, \ldots, k$. We formulate the model as

$$\bar{\mathbf{S}}_i = \sum_{l=1}^{k} \omega_{il}\mathbf{B}_l \tag{4.19}$$

where $\omega_{il} \in \mathbb{R}$ is the deformation weight for base l at frame i. A perfect rigid object corresponds to the situation of $k = 1$ and $\omega_{il} = 1$. Suppose all image features are registered to their centroid in each frame, then we have the following formulation under orthographic projection.

$$\begin{aligned} \mathbf{W}_i &= [\bar{\mathbf{x}}_{i1}, \bar{\mathbf{x}}_{i2}, \ldots, \bar{\mathbf{x}}_{in}] = \mathbf{R}_{Ai}\left(\sum_{l=1}^{k} \omega_{il}\mathbf{B}_l\right) \\ &= [\omega_{i1}\mathbf{R}_{Ai}, \ldots, \omega_{ik}\mathbf{R}_{Ai}] \begin{bmatrix} \mathbf{B}_1 \\ \vdots \\ \mathbf{B}_k \end{bmatrix}, \quad \forall i = 1, \ldots, m \end{aligned} \tag{4.20}$$

where $\mathbf{R}_{Ai}$ stands for the first two rows of the rotation matrix corresponding to the ith frame. Then we can obtain the factorization expression of the tracking matrix by stacking all equations in (4.20) frame by frame.

$$\underbrace{\begin{bmatrix} \bar{\mathbf{x}}_{11} & \cdots & \bar{\mathbf{x}}_{1n} \\ \vdots & \ddots & \vdots \\ \bar{\mathbf{x}}_{m1} & \cdots & \bar{\mathbf{x}}_{mn} \end{bmatrix}}_{\mathbf{W}_{2m\times n}} = \underbrace{\begin{bmatrix} \omega_{11}\mathbf{R}_{A1} & \cdots & \omega_{1k}\mathbf{R}_{A1} \\ \vdots & \ddots & \vdots \\ \omega_{m1}\mathbf{R}_{Am} & \cdots & \omega_{mk}\mathbf{R}_{Am} \end{bmatrix}}_{\mathbf{M}_{2m\times 3k}} \underbrace{\begin{bmatrix} \mathbf{B}_1 \\ \vdots \\ \mathbf{B}_k \end{bmatrix}}_{\bar{\mathbf{B}}_{3k\times n}} \tag{4.21}$$

The above equation can be written in short as $\mathbf{W} = \mathbf{M}\bar{\mathbf{B}}$, which is similar as the rigid factorization (4.7). The only difference lies in the form and dimension of the motion and shape matrices. From the right side of (4.21), it is easy to find that the rank of the nonrigid tracking matrix is at most $3k$ (usually $2m$ and n are both larger than $3k$). The goal of nonrigid factorization is to recover the motion and deformation structure corresponding to each frame.

4.4.2 Nonrigid Factorization Under Affine Models

Following the idea of rigid factorization, we perform SVD decomposition on the nonrigid tracking matrix and impose the rank-$3k$ constraint, $\mathbf{W}$ can be factorized into a $2m \times 3k$ matrix $\tilde{\mathbf{M}}$ and a $3k \times n$ matrix $\tilde{\mathbf{B}}$. However, the decomposition is not unique as any nonsingular linear transformation matrix $\mathbf{H} \in \mathbb{R}^{3k\times 3k}$ can be inserted into the factorization which leads to an alternative factorization $\mathbf{W} = (\tilde{\mathbf{M}}\mathbf{H})(\mathbf{H}^{-1}\tilde{\bar{\mathbf{B}}})$. If we have a transformation matrix $\mathbf{H}$ that can resolve the affine ambiguity and upgrade the solution to Euclidean space, the shape bases are then easily recovered from $\mathbf{B} = \mathbf{H}^{-1}\tilde{\mathbf{B}}$, while the rotation matrix $\mathbf{R}_{Ai}$ and the weighting coefficient ω_{ij} can be decomposed from $\mathbf{M} = \tilde{\mathbf{M}}\mathbf{H}$ by Procrustes analysis [4, 6, 37]. Similar to the rigid situation, the upgrading matrix is usually recovered by application of metric constraint to the motion matrix. In the following section, we will briefly review some typical methods to deal with this problem.

4.4.2.1 Metric Constraints

To recover the upgrading matrix, many researchers apply metric constraint on the rotation matrix. Bregler *et al.* [6] first introduced the nonrigid factorization framework and proposed a sub-block factorization algorithm to decompose each two-row sub-block in $\tilde{\mathbf{M}}$ into the rotation matrix $\mathbf{R}_{Ai}$ and deformation weights ω_{il}.

In (4.21), each two-row sub-block is given as

$$\mathbf{M}_i = [\omega_{i1}\mathbf{R}_{Ai}, \ldots, \omega_{ik}\mathbf{R}_{Ai}] \tag{4.22}$$

which can be rearranged as a $k \times 6$ matrix $\mathbf{M}'_i$ as follows.

$$\mathbf{M}'_i = \begin{bmatrix} \omega_{i1}\mathbf{r}_{1i}^T \mid \omega_{i1}\mathbf{r}_{2i}^T \\ \vdots \quad\quad \vdots \\ \omega_{ik}\mathbf{r}_{1i}^T \mid \omega_{ik}\mathbf{r}_{2i}^T \end{bmatrix}_{k\times 6} = \begin{bmatrix} \omega_{i1} \\ \vdots \\ \omega_{ik} \end{bmatrix} \left[\mathbf{r}_{1i}^T | \mathbf{r}_{2i}^T\right] \tag{4.23}$$

where $\mathbf{r}_{1i}^T$ and $\mathbf{r}_{2i}^T$ are the first and second rows of the rotation matrix. Clearly, the deformation weight ω_{il} and rotation matrix can be easily derived via the SVD factorization of $\mathbf{M}'_i$ with rank-1 constraint. However, such recovered rotation matrix

$$\tilde{\mathbf{R}}_{Ai} = \begin{bmatrix} \mathbf{r}_{1i}^T \\ \mathbf{r}_{2i}^T \end{bmatrix}$$

is usually not an orthonormal matrix, thus an orthonormality process is enforced to find a transformation matrix $\mathbf{H}$ using the metric constraint [35] so that we can upgrade the solution to Euclidean space.

For recovery of the upgrading matrix, Xiao *et al.* [44] presented a block computation method via metric constraint. Suppose the lth column triples of $\mathbf{H}$ is $\mathbf{H}_l$, $l = 1, \ldots, k$, which is independent of each other since $\mathbf{H}$ is nonsingular. Then we have

$$\tilde{\mathbf{M}}\mathbf{H}_l = \begin{bmatrix} \omega_{1l}\mathbf{R}_{A1} \\ \vdots \\ \omega_{ml}\mathbf{R}_{Am} \end{bmatrix} \tag{4.24}$$

and

$$\tilde{\mathbf{M}}\mathbf{H}_l\mathbf{H}_l^T\tilde{\mathbf{M}}^T = \begin{bmatrix} \omega_{1l}^2\mathbf{R}_{A1}\mathbf{R}_{A1}^T & * & \cdots & * \\ * & \omega_{2l}^2\mathbf{R}_{A2}\mathbf{R}_{A2}^T & \cdots & * \\ \vdots & \vdots & \ddots & \vdots \\ * & * & \cdots & \omega_{ml}^2\mathbf{R}_{Am}\mathbf{R}_{Am}^T \end{bmatrix} \tag{4.25}$$

where '$*$' stands for nonzero entries. Let $\mathbf{Q}_l = \mathbf{H}_l\mathbf{H}_l^T$, which is a $3k \times 3k$ symmetric matrix with $\frac{1}{2}k(9k+1)$ unknowns. In any diagonal block in (4.25), $\mathbf{R}_{Ai}$ is an orthonormal rotation matrix, thus we have

$$\tilde{\mathbf{M}}_{2i-1:2i}\mathbf{Q}_l\tilde{\mathbf{M}}_{2i-1:2i}^T = \omega_{il}^2 \begin{bmatrix} 1 & 0 \\ 0 & 1 \end{bmatrix} \tag{4.26}$$

where $\tilde{\mathbf{M}}_{2i-1:2i}$ stands for the ith two-row of $\tilde{\mathbf{M}}$. Thus we have the following linear constraints on $\mathbf{Q}_l$.

$$\begin{cases} \tilde{\mathbf{M}}_{2i-1}\mathbf{Q}_l\tilde{\mathbf{M}}_{2i}^T = 0 \\ \tilde{\mathbf{M}}_{2i-1}\mathbf{Q}_l\tilde{\mathbf{M}}_{2i-1}^T - \tilde{\mathbf{M}}_{2i}\mathbf{Q}_l\tilde{\mathbf{M}}_{2i}^T = 0 \end{cases} \quad \forall i = 1, \ldots, m \tag{4.27}$$

4

Therefore, $\mathbf{Q}_l$ may be computed linearly via least squares from (4.27) if we have sufficient frames, then $\mathbf{H}_l$ is recovered from $\mathbf{Q}_l$ via Cholesky decomposition.

In sub-block factorization, it is assumed that all configurations concerning the camera motion and deformation weights are contained in the initially rank-$3k$ factorized matrix $\tilde{\mathbf{M}}$. While the initial decomposition may factorize the components of $\tilde{\mathbf{M}}$ and $\tilde{\mathbf{B}}$ randomly [4, 10], which may result in a bad estimation of the rotation matrix. To overcome the limitation of sub-block factorization, Torresani *et al.* [37] proposed a tri-linear approach to solve $\mathbf{B}_l$, ω_{il}, and $\mathbf{R}_{Ai}$ alternatively by iteration. They define equation (4.20) by minimizing the following cost function.

$$f(\mathbf{B}_l, \omega_{il}, \mathbf{R}_{Ai}) = \left\| \mathbf{W}_i - \mathbf{R}_{Ai}\left(\sum_{l=1}^{k} \omega_{il}\mathbf{B}_l\right)\right\|_F^2, \quad \forall i = 1, \ldots, m \tag{4.28}$$

The algorithm is initialized by rigid assumption. They perform a rigid factorization [35] on the nonrigid tracking matrix and obtain an average shape (mean shape) matrix $\tilde{\mathbf{S}}_{rig}$ and a rigid rotation matrix $\tilde{\mathbf{R}}_{Ai}$, $i = 1, \ldots, m$, for each frame. The deformation weights are initialized randomly as ω_{il}. Then iteratively perform the following three steps.

1. Estimate the shape bases $\bar{\mathbf{B}}$ from $\tilde{\mathbf{R}}_{Ai}$ and ω_{il};
2. Update the deformation weight ω_{il} from $\bar{\mathbf{B}}$ and $\tilde{\mathbf{R}}_{Ai}$;
3. Update the rotation matrix $\tilde{\mathbf{R}}_{Ai}$ from $\bar{\mathbf{B}}$ and ω_{il}.

The procedure is simpler than general nonlinear method and may converge to a proper solution. However, the algorithm does not preserve the replicated block structure of the motion matrix (4.21) during iterations [10]. Similar to tri-linear technique, Wang *et al.* [41] proposed a rotation constrained power factorization technique by combining the orthonormality of the rotation matrix into power factorization algorithm. In addition, Brand [4] proposed a flexible factorization technique to compute the upgrading matrix and recover the motion parameters and deformation weights using an alternative orthonormal decomposition algorithm.

4.4.2.2 Basis Constraints

One main problem of using metric constraint to recover the upgrading matrix lies in its ambiguity. Given the same tracking data, different motion and deformable shapes may be found, since any nonsingular linear transformation matrix can be inserted into the factorization process and will therefore lead to different sets of eligible shape bases. On the other hand, when we use the constraints (4.27) to recover the matrix $\mathbf{Q}_l$, it appears that if we have enough features and frames, the upgrading matrix can be solved linearly by stacking all the constraints in (4.27). Unfortunately, only the rotation constraints may be insufficient when the object deforms at varying speed, since most of the constraints are redundant. Xiao *et al.* [45] proposed a basis constraint to solve this ambiguity.

The main idea is based on the assumption that there exists k frames in the sequence which include independent shapes that can be treated as a set of bases. Suppose the first k

frames are independent of each other, then their corresponding weighting coefficients can be set as

$$\omega_{il} = \begin{cases} 1 & \text{if } i, l = 1, \ldots, k \text{ and } i = l \\ 0 & \text{if } i, l = 1, \ldots, k \text{ and } i \neq l \end{cases} \tag{4.29}$$

Let us define $\Omega = \{(i, j) | i = 1, \ldots, k, j = 1, \ldots, m, i \neq l\}$, then from (4.25) we can obtain following basis constraint.

$$\begin{cases} \tilde{\mathbf{M}}_{2i-1}\mathbf{Q}_l\tilde{\mathbf{M}}_{2j-1}^T = \begin{cases} 1, & i = j = l \\ 0, & (i, j) \in \Omega \end{cases} \\ \tilde{\mathbf{M}}_{2i}\mathbf{Q}_l\tilde{\mathbf{M}}_{2j}^T = \begin{cases} 1, & i = j = l \\ 0, & (i, j) \in \Omega \end{cases} \\ \tilde{\mathbf{M}}_{2i-1}\mathbf{Q}_l\tilde{\mathbf{M}}_{2j}^T = \begin{cases} 0, & i = j = l \\ 0, & (i, j) \in \Omega \end{cases} \\ \tilde{\mathbf{M}}_{2i}\mathbf{Q}_l\tilde{\mathbf{M}}_{2j-1}^T = \begin{cases} 0, & i = j = l \\ 0, & (i, j) \in \Omega \end{cases} \end{cases} \tag{4.30}$$

Altogether we have $4m(k-1)$ linear basis constraints. Using both the metric constraints (4.27) and basis constraints (4.30), Xiao *et al.* [45] derived a linear close form solution to nonrigid factorization by dividing the problem into k linear systems. However, the method deals with each columns triples $\mathbf{M}_i$ separately, thus the repetitive block structure of the entire motion matrix is not observed during computation. The solution is dependant on the selection of shape bases that were treated as prior information of the deformation and such a selection may be difficult in some situations. Following this idea, Brand [5] proposed a modified approach based on the deviation of the solution from metric constraints. They apply a weak constraint on the independent shape bases.

4.4.3 Nonrigid Factorization Under Perspective Projection

In this section, we will discuss the perspective reconstruction method proposed in [45]. From perspective projection (4.1), we can obtain the following equation by stacking the projection of each frame.

$$\dot{\mathbf{W}} = \begin{bmatrix} \lambda_{11}\mathbf{x}_{11} & \cdots & \lambda_{1n}\mathbf{x}_{1n} \\ \vdots & \ddots & \vdots \\ \lambda_{m1}\mathbf{x}_{m1} & \cdots & \lambda_{mn}\mathbf{x}_{mn} \end{bmatrix} = \begin{bmatrix} \mathbf{P}_1\mathbf{S}_1 \\ \vdots \\ \mathbf{P}_m\mathbf{S}_m \end{bmatrix} \tag{4.31}$$

where $\mathbf{S}_i$ is a $4 \times n$ matrix that denotes the 3D structure corresponding to the ith frame in homogeneous form, $\mathbf{1}^T$ is a n-vector with unit entries. For rigid objects, the shape

4

does not change with time, thus we have $\mathbf{S}_1 = \cdots = \mathbf{S}_m$. Following Bregler's deformation model (4.19), we have

$$\mathbf{P}_i\mathbf{S}_i = \sum_{l=1}^{k} \left(\omega_{il}\mathbf{P}_i^{(1:3)}\mathbf{B}_l\right) + \mathbf{P}_i^{(4)}\mathbf{1}^T \tag{4.32}$$

where $\mathbf{P}_i^{(1:3)}$ and $\mathbf{P}_i^{(4)}$ denote the first three columns and the last column of $\mathbf{P}_i$ respectively. From (4.31) and (4.32), we obtain the following expression for perspective nonrigid factorization.

$$\dot{\mathbf{W}} = \underbrace{\begin{bmatrix} \omega_{11}\mathbf{P}_1^{(1:3)} & \cdots & \omega_{1k}\mathbf{P}_1^{(1:3)} & \mathbf{P}_1^{(4)} \\ \vdots & \ddots & \vdots & \vdots \\ \omega_{m1}\mathbf{P}_m^{(1:3)} & \cdots & \omega_{mk}\mathbf{P}_m^{(1:3)} & \mathbf{P}_m^{(4)} \end{bmatrix}}_{\mathbf{M}_{3m\times(3k+1)}} \underbrace{\begin{bmatrix} \mathbf{B}_1 \\ \vdots \\ \mathbf{B}_k \\ \mathbf{1}^T \end{bmatrix}}_{\mathbf{B}_{(3k+1)\times n}} \tag{4.33}$$

where $\mathbf{M} \in \mathbb{R}^{3m\times(3k+1)}$ and $\mathbf{B} \in \mathbb{R}^{(3k+1)\times n}$ are the motion matrix and shape bases. All nonrigid structures of one object share the same shape bases, and both $\mathbf{M}$ and $\mathbf{B}$ are of full rank. Therefore, the rank of the scale weighted tracking matrix is no more than $\min((3k+1), 3m, n)$. In practice, the point and frame numbers are usually large than the shape bases number, so the rank of $\dot{\mathbf{W}}$ is at most $3k+1$. This is consistent with that of rigid factorization, where $k = 1$ and the rank is no more than 4.

Suppose the projective depth scales in $\dot{\mathbf{W}}$ are available, then a projective solution $\tilde{\mathbf{M}}$ and $\tilde{\mathbf{B}}$ can be obtained by SVD factorization on the weighted tracking matrix with rank-$(3k+1)$ constraint. Obviously, the solution is defined up to a nonsingular transformation matrix $\mathbf{H} \in \mathbb{R}^{(3k+1)\times(3k+1)}$. Similar to the rigid case, we adopt the metric constraints and basis constraints to compute the upgrading matrix, then the Euclidean motion parameters and deformation structures are recovered from $\tilde{\mathbf{M}}$ and $\tilde{\mathbf{B}}$. Please refer to [45] for computation details.

As for the recovery of the projective depths, a similar iteration method as in [15] and [24] based on the rank constraint on $\dot{\mathbf{W}}$ is adopted, which starts with weak perspective assumption and sets $\lambda_{ij} = 1$. The depth scales are optimized iteratively by minimizing the following cost function

$$J(\lambda_{ij}) = \min \|\dot{\mathbf{W}} - \tilde{\mathbf{M}}\tilde{\mathbf{B}}\|_F^2 \tag{4.34}$$

The minimizing process is achieved iteratively by first factorizing $\dot{\mathbf{W}}$ into $\tilde{\mathbf{M}}\tilde{\mathbf{B}}$ with the given depth scales and then upgrading the depth scales by back projection. In deformation case, the rank of the tracking matrix is $3k+1$ and the dimension of $\tilde{\mathbf{M}}$ and $\tilde{\mathbf{B}}$ follows (4.33). In order to avoid trivial solutions $\lambda_{ij} = 0$, we apply the following constraints in alternative steps so that the depth scales of all points in any frame or a single point in all images have unit norms, such that the minimization of (4.34) is converted to a simple eigenvalue problem [45].

4.5 Factorization of Multi-Body and Articulated Objects

The nonrigid objects discussed above have regular deformations. In real world, there are some other nonrigid forms, scenarios with independent moving objects, etc. In these cases, it is still possible to formulate the structure and motion recovery under the factorization framework.

4.5.1 Multi-Body Factorization

The extension of structure and motion factorization to multiple moving objects was proposed by Costeira and Kanade [9] based on orthographic projection. When there are several moving objects in the scene, as shown in Fig. 4.5, the tracking matrix $\mathbf{W}$ contains features of different objects originated from the motions. Suppose the objects are segmented and the features are sorted according to the object. Then we can permute the columns of $\mathbf{W}$ and rewrite it into the canonical form

$$\mathbf{W}^* \equiv [\mathbf{W}_1|\mathbf{W}_2|\cdots|\mathbf{W}_k] \tag{4.35}$$

where $\mathbf{W}_l$ is the tracking matrix of lth object and k is the number of object. Suppose each submatrix is factorized as

$$\mathbf{W}_l = \mathbf{U}_l\Sigma_l\mathbf{V}_l^T = \mathbf{M}_l\mathbf{S}_l \tag{4.36}$$

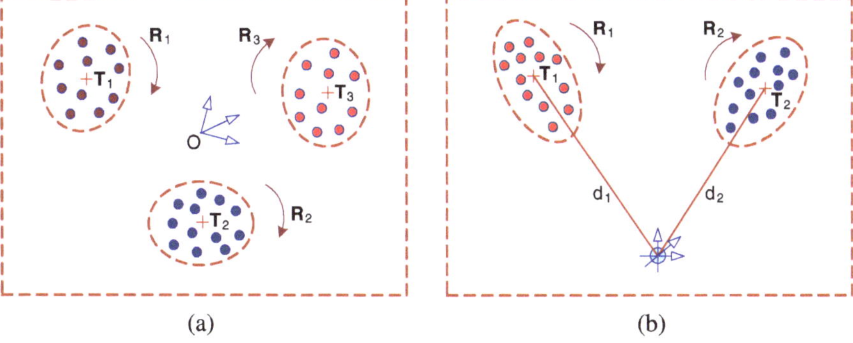

Fig. 4.5 (**a**) Three independent moving objects in the scene, where $\mathbf{R}_1, \mathbf{R}_2, \mathbf{R}_3$ and $\mathbf{T}_1, \mathbf{T}_2, \mathbf{T}_3$ are the corresponding rotation matrices and centroids of the objects; (**b**) An articulated object with one joint and two movable arms, where d_1 and d_2 are the distances between the joint and the centroids of the two groups of features

The factorization is defined up to upgrading matrix. After upgrading to Euclidean space, the motion part $\mathbf{M}_l$ is a $2m \times 4$ matrix of the form

$$\begin{bmatrix} \mathbf{r}_1^T & \mathbf{t}_u \\ \mathbf{r}_2^T & \mathbf{t}_v \end{bmatrix}$$

the shape part $\mathbf{S}_l$ is a $4 \times n$ matrix in homogeneous form, and the rank of $\mathbf{W}_l$ is at most 4. This is different from rigid factorization where the tracking matrix is of rank 3. The variation in rank is due to the fact images of multiple moving objects can not be registered to a common centroid. Thus the translation part in (4.2) can not be eliminated. Simply, the factorization of $\mathbf{W}^*$ can be written as

$$\mathbf{W}^* = \mathbf{U}^* \Sigma^* \mathbf{V}^{*T} = \mathbf{M}^* \mathbf{S}^* \tag{4.37}$$

where

$$\mathbf{U}^* = [\mathbf{U}_1 | \mathbf{U}_2 | \cdots | \mathbf{U}_k], \qquad \mathbf{M}^* = [\mathbf{M}_1 | \mathbf{M}_2 | \cdots | \mathbf{M}_k]$$

$$\Sigma^* = \begin{bmatrix} \Sigma_1 & \mathbf{0} & \cdots & \mathbf{0} \\ \mathbf{0} & \Sigma_2 & \cdots & \mathbf{0} \\ \vdots & \vdots & \ddots & \vdots \\ \mathbf{0} & \mathbf{0} & \cdots & \Sigma_k \end{bmatrix}, \qquad \mathbf{V}^* = \begin{bmatrix} \mathbf{V}_1 & \mathbf{0} & \cdots & \mathbf{0} \\ \mathbf{0} & \mathbf{V}_2 & \cdots & \mathbf{0} \\ \vdots & \vdots & \ddots & \vdots \\ \mathbf{0} & \mathbf{0} & \cdots & \mathbf{V}_k \end{bmatrix}$$

$$\mathbf{S}^* = \begin{bmatrix} \mathbf{S}_1 & \mathbf{0} & \cdots & \mathbf{0} \\ \mathbf{0} & \mathbf{S}_2 & \cdots & \mathbf{0} \\ \vdots & \vdots & \ddots & \vdots \\ \mathbf{0} & \mathbf{0} & \cdots & \mathbf{S}_k \end{bmatrix}$$

It is clear from (4.35) that $\mathbf{W}^*$ is at most of rank $4k$. In non-degenerate case, we can perform SVD factorization of the tracking matrix and truncating it to rank $4k$. Then upgrade the solution to Euclidean space via metric constraint as discussed in the last section.

In reality, a general tracking matrix $\mathbf{W}$ is a mixture of features from different objects, and we do not know which feature belongs to which object. The key task here is to segment different objects by observing the tracking matrix. Costeira and Kanade [9] first introduced a shape interaction matrix to perform this task.

Suppose the rank-$4k$ SVD factorization of $\mathbf{W}$ is $\mathbf{W} = \mathbf{U}\Sigma\mathbf{V}^T$. The shape interaction matrix is defined as

$$\mathbf{Q} \equiv \mathbf{V}\mathbf{V}^T \tag{4.38}$$

From the definition, we find that swapping the columns of $\mathbf{W}$ does not change the set of entries in $\mathbf{Q}$. Specifically, permuting the columns i and j of $\mathbf{W}$ is equivalent to swapping the columns j and i of $\mathbf{V}^T$, which results in simultaneous swapping of the columns i and j and rows i and j in $\mathbf{Q}$. For the canonical form of tracking matrix $\mathbf{W}^*$, the corresponding interaction matrix becomes

$$\mathbf{Q}^* = \mathbf{V}^*\mathbf{V}^{*T} = \begin{bmatrix} \mathbf{S}_1^T\Lambda_1^{-1}\mathbf{S}_1 & \mathbf{0} & \cdots & \mathbf{0} \\ \mathbf{0} & \mathbf{S}_2^T\Lambda_2^{-1}\mathbf{S}_2 & \cdots & \mathbf{0} \\ \vdots & \vdots & \ddots & \vdots \\ \mathbf{0} & \mathbf{0} & \cdots & \mathbf{S}_k^T\Lambda_k^{-1}\mathbf{S}_k \end{bmatrix} \tag{4.39}$$

where Λ_l is a 4×4 matrix of the moment of inertia of each object. For noise free data, the entries Q_{ij}^* of $\mathbf{Q}^*$ have the following property

$$Q_{ij}^* \begin{cases} \neq 0, & \text{if the } i\text{th and } j\text{th points belong to the same objects} \\ = 0, & \text{if the } i\text{th and } j\text{th points belong to different objects} \end{cases} \tag{4.40}$$

It is clear that a general shape interaction matrix $\mathbf{Q}$ has the same set of entries as $\mathbf{Q}^*$. Each entry Q_{ij} can be interpreted as a measure of interaction between features i and j. If the two points belong to different objects, $Q_{ij} = 0$; otherwise, they belong to the same object. The problem is now reduced to sorting the entries of matrix $\mathbf{Q}$ by swapping pairs of rows and columns until different moving objects are separated, i.e. $\mathbf{Q}$ becomes block diagonal. Figure 4.6 illustrates the segmentation process of two moving objects.

In the presence of noise, the sorting process becomes difficult since the entry Q_{ij} no longer vanishes even the features are from different objects. Costeira and Kanade [9] proposed an iterative minimization procedure to minimize the total energy of $\mathbf{Q}$. Kanatani [20] proposed a robust segmentation technique by dimension correction, model selection, and least-median fitting. In a more recent study, Ozden *et al.* [27] discussed many theoretical and practical issues in multi-body structure from motion. Some further studies can be found in [14, 34, 49, 51].

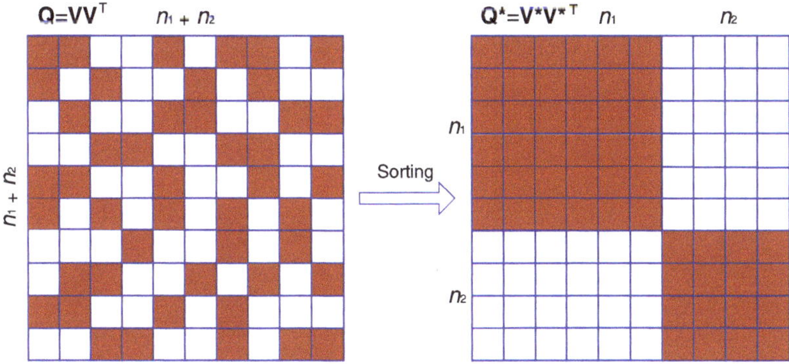

Fig. 4.6 Example of shape interaction matrix obtained from two rigid objects. $\mathbf{Q}$ is the shape interaction matrix derived from a general tracking matrix $\mathbf{W}$ before segmentation. $\mathbf{Q}^*$ is the matrix after sorting the trajectories into two clusters of independent motions, which corresponds to the canonical tracking matrix $\mathbf{W}^*$. n_1 and n_2 are the feature numbers of the two objects

Fig. 4.7 Some examples of articulated objects linked by axes and universal joints

4.5.2 Articulated Factorization

An articulated object consists of multiple parts that are linked by a joint or an axis, such as a robot, arms, human head and torso, etc. Figure 4.7 shows some examples of articulated objects.

As discussed in the last section, the rank of the tracking matrix for each moving object is 4. For two independent moving objects, the rank of the tracking matrix will be 8 in non-degenerate case. However, when the two objects are linked together, the subspace representing the tracking matrix will decrease by one or two dimensions depending on the type of the joint.

Without loss of generality, we assume there are two linked parts as shown in Fig. 4.5. Suppose the world coordinate system is selected on the joint, and the object to camera motion is expressed with respect to the coordinate system of the first part. Then the structure of the other part in ith frame can be expressed relative to the first part as

$$\bar{\mathbf{S}}_{2i} = \mathbf{O}_i \bar{\mathbf{S}}_2 \tag{4.41}$$

where $\mathbf{O}_i$ is a rotation matrix, $\bar{\mathbf{S}}$ is shape matrix in non-homogeneous form. Then the projection of ith frame can be expressed by

$$\mathbf{W}_i = [\mathbf{R}_i | \mathbf{T}_i]\mathbf{S}_1 + [\mathbf{R}_i \mathbf{O}_i | \mathbf{T}_i]\mathbf{S}_2 \tag{4.42}$$

Therefore, the factorization of the tracking matrix can be formulated as

$$\mathbf{W} = \begin{bmatrix} \mathbf{R}_1 & | & \mathbf{T}_1 & | & \mathbf{R}_1\mathbf{O}_1 & | & \mathbf{T}_1 \\ \vdots & & \vdots & & \vdots & & \vdots \\ \mathbf{R}_m & | & \mathbf{T}_m & | & \mathbf{R}_m\mathbf{O}_m & | & \mathbf{T}_m \end{bmatrix} \begin{bmatrix} \mathbf{S}_1 & \mathbf{0} \\ \mathbf{0} & \mathbf{S}_2 \end{bmatrix} \tag{4.43}$$

For joint link, $\mathbf{O}_i$ is a general 3D rotation matrix. From the right side of (4.43) we can see that the rank of $\mathbf{W}$ is at most 7. For an axis joint, we can select the Z-axis of the world frame as the rotation axis. Then the rotation matrix becomes

$$\mathbf{O}_i = \begin{bmatrix} \cos\theta_i & \sin\theta_i & 0 \\ -\sin\theta_i & \cos\theta_i & 0 \\ 0 & 0 & 1 \end{bmatrix} \tag{4.44}$$

Therefore, the last columns of $\mathbf{R}_i$ and $\mathbf{R}_i\mathbf{O}_i$ are identical, and the rank of the tracking matrix is reduced to 6.

Tresadern and Reid [38] first proposed the articulation constraint and imposed it to factorization and self-calibration. Yan and Pollefeys [47] presented a good study on articulated nonrigid factorization and proposed a method to find the axes and joints in articulated motion.

4.6 Closure Remarks

4.6.1 Conclusion

In the chapter, we briefly reviewed the structure and motion factorization algorithm under different projection models for both rigid and nonrigid objects. When the tracking data of a sequence of images is available, the factorization algorithm offers several advantages over other techniques, since it explores all data uniformly, thus good robustness and accuracy may be achieved.

All methods presented in the chapter are based on the SVD factorization with rank constraint to find the initial solution in affine or projective space. The solution is defined up to an upgrading matrix so that we can stratify the solution from affine or projective space to Euclidean space. The metric constraints are used to recover the upgrading matrix by exploring the orthonormality of the rotation matrix. The approach works well in rigid factorization, whereas in nonrigid factorization, further basis constraints are employed. Besides, the deformation weights and motion parameters are strongly coupled in nonrigid situations, a special care should be taken to have these parameters decoupled.

When perspective camera model is considered, the tracking matrix is weighted by unknown depth scales. Most methods adopt an iterative scheme to recover these scales starting from affine assumption. However, there is no evidence that such iteration will converge to a set of correct projective depths. As suggested in Oliensis and Hartley [26], no iteration has shown to converge sensibly.

Feature tracking is a difficult problem especially for deformation objects. Most methods assume the tracking data is available, while in practice, some features may be unavailable in certain frames due to self occlusion, illumination changes, or tracking failures. It is hard to deal with missing data for SVD-based methods. We will show some recent developments of the algorithm in the coming chapters.

4.6.2 Review Questions

1. *Rigid factorization.* Determine the rigid factorization expressions under both affine assumption and perspective projection model. What is the difference between the two

4

expressions? How to use the metric constraint to calibrate the cameras and recover the upgrading matrix?

2. *Nonrigid factorization*. Give the nonrigid factorization expressions under affine projection model. What is the difference between nonrigid and rigid factorization? How to use metric constraints and basic constraints to recover the upgrading matrix?
3. *Multi-body and articulated objects*. Illustrate the rank constraint of multi-body tracking matrix. How to use the shape interaction matrix to segment different objects? Derive the rank constraint of a universal joint and axis linked object. Estimate the tracking matrix rank of an object linked by multiple joints and axes.

References

1. Aanæs, H., Fisker, R., Åström, K., Carstensen, J.M.: Robust factorization. IEEE Trans. Pattern Anal. Mach. Intell. **24**(9), 1215–1225 (2002)
2. Anandan, P., Irani, M.: Factorization with uncertainty. Int. J. Comput. Vis. **49**(2–3), 101–116 (2002)
3. Bascle, B., Blake, A.: Separability of pose and expression in facial tracing and animation. In: Proc. of the International Conference on Computer Vision, pp. 323–328 (1998)
4. Brand, M.: Morphable 3D models from video. In: Proc. of IEEE Conference on Computer Vision and Pattern Recognition, vol. 2, pp. 456–463 (2001)
5. Brand, M.: A direct method for 3D factorization of nonrigid motion observed in 2D. In: Proc. of IEEE Conference on Computer Vision and Pattern Recognition, vol. 2, pp. 122–128 (2005)
6. Bregler, C., Hertzmann, A., Biermann, H.: Recovering non-rigid 3D shape from image streams. In: Proc. of IEEE Conference on Computer Vision and Pattern Recognition, vol. 2, pp. 690–696 (2000)
7. Buchanan, A.M., Fitzgibbon, A.W.: Damped newton algorithms for matrix factorization with missing data. In: Proc. of IEEE Conference on Computer Vision and Pattern Recognition, vol. 2, pp. 316–322 (2005)
8. Christy, S., Horaud, R.: Euclidean shape and motion from multiple perspective views by affine iterations. IEEE Trans. Pattern Anal. Mach. Intell. **18**(11), 1098–1104 (1996)
9. Costeira, J., Kanade, T.: A multibody factorization method for independent moving objects. Int. J. Comput. Vis. **29**(3), 159–179 (1998)
10. Del Bue, A.: Deformable 3-D modelling from uncalibrated video sequences. Ph.D. Thesis, Queen Mary, University of London (2007)
11. Del Bue, A., Agapito, L.: Non-rigid stereo factorization. Int. J. Comput. Vis. **66**(2), 193–207 (2006)
12. Del Bue, A., Lladó, X., de Agapito, L.: Non-rigid metric shape and motion recovery from uncalibrated images using priors. In: Proc. of IEEE Conference on Computer Vision and Pattern Recognition, vol. 1, pp. 1191–1198 (2006)
13. Del Bue, A., Smeraldi, F., Agapito, L.: Non-rigid structure from motion using nonparametric tracking and non-linear optimization. In: Proc. of IEEE Workshop in Articulated and Nonrigid Motion, pp. 8–15 (2004)
14. Gruber, A., Weiss, Y.: Multibody factorization with uncertainty and missing data using the EM algorithm. In: Proc. of IEEE Conference on Computer Vision and Pattern Recognition, vol. 1, pp. 707–714 (2004)
15. Han, M., Kanade, T.: Creating 3D models with uncalibrated cameras. In: Proc. of IEEE Computer Society Workshop on the Application of Computer Vision (2000)
16. Hartley, R., Vidal, R.: Perspective nonrigid shape and motion recovery. In: Proc. of European Conference on Computer Vision. Lecture Notes in Computer Science, vol. 5302, pp. 276–289. Springer, Berlin (2008)

17. Hartley, R.I., Zisserman, A.: Multiple View Geometry in Computer Vision, 2nd edn. Cambridge University Press, Cambridge (2004). ISBN: 0521540518
18. Heyden, A., Berthilsson, R., Sparr, G.: An iterative factorization method for projective structure and motion from image sequences. Image Vis. Comput. **17**(13), 981–991 (1999)
19. Kanade, T., Morris, D.D.: Factorization methods for structure from motion. Philos. Trans. R. Soc. Lond. Ser. A **356**, 1153–1173 (2001)
20. Kanatani, K.: Motion segmentation by subspace separation and model selection. In: Proc. of International Conference on Computer Vision, vol. 2 (2001)
21. Li, T., Kallem, V., Singaraju, D., Vidal, R.: Projective factorization of multiple rigid-body motions. In: Proc. of IEEE Conference on Computer Vision and Pattern Recognition (2007)
22. Lowe, D.G.: Object recognition from local scale-invariant features. In: Proc. of the International Conference on Computer Vision, vol. 2, pp. 1150–1157 (1999)
23. Lucas, B.D., Kanade, T.: An iterative image registration technique with an application to stereo vision. In: Proc. of International Joint Conference on Artificial Intelligence (IJCAI), pp. 674–679 (1981)
24. Mahamud, S., Hebert, M.: Iterative projective reconstruction from multiple views. In: Proc. of IEEE Conference on Computer Vision and Pattern Recognition, vol. 2, pp. 430–437 (2000)
25. Okatani, T., Deguchi, K.: On the Wiberg algorithm for matrix factorization in the presence of missing components. Int. J. Comput. Vis. **72**(3), 329–337 (2007)
26. Oliensis, J., Hartley, R.: Iterative extensions of the Sturm/Triggs algorithm: Convergence and nonconvergence. IEEE Trans. Pattern Anal. Mach. Intell. **29**(12), 2217–2233 (2007)
27. Ozden, K.E., Schindler, K., Van Gool, L.: Multibody structure-from-motion in practice. IEEE Trans. Pattern Anal. Mach. Intell. **32**(6), 1134–1141 (2010)
28. Poelman, C., Kanade, T.: A paraperspective factorization method for shape and motion recovery. IEEE Trans. Pattern Anal. Mach. Intell. **19**(3), 206–218 (1997)
29. Quan, L.: Self-calibration of an affine camera from multiple views. Int. J. Comput. Vis. **19**(1), 93–105 (1996)
30. Quan, L., Kanade, T.: A factorization method for affine structure from line correspondences. In: Proc. of IEEE Conference on Computer Vision and Pattern Recognition, pp. 803–808 (1996)
31. Rabaud, V., Belongie, S.: Re-thinking non-rigid structure from motion. In: Proc. of IEEE Conference on Computer Vision and Pattern Recognition (2008)
32. Shi, J., Tomasi, C.: Good features to track. In: Proc. of IEEE Conference on Computer Vision and Pattern Recognition, pp. 593–600 (1994)
33. Sturm, P.F., Triggs, B.: A factorization based algorithm for multi-image projective structure and motion. In: Proc. of European Conference on Computer Vision, vol. 2, pp. 709–720 (1996)
34. Sugaya, Y., Kanatani, K.: Outlier removal for motion tracking by subspace separation. IEICE Trans. Inf. Syst. **86**, 1095–1102 (2003)
35. Tomasi, C., Kanade, T.: Shape and motion from image streams under orthography: A factorization method. Int. J. Comput. Vis. **9**(2), 137–154 (1992)
36. Torresani, L., Hertzmann, A., Bregler, C.: Nonrigid structure-from-motion: Estimating shape and motion with hierarchical priors. IEEE Trans. Pattern Anal. Mach. Intell. **30**(5), 878–892 (2008)
37. Torresani, L., Yang, D.B., Alexander, E.J., Bregler, C.: Tracking and modeling non-rigid objects with rank constraints. In: Proc. of IEEE Conference on Computer Vision and Pattern Recognition, vol. 1, pp. 493–500 (2001)
38. Tresadern, P., Reid, I.: Articulated structure from motion by factorization. In: Proc. of the IEEE Conference on Computer Vision and Pattern Recognition, vol. 2, pp. 1110–1115 (2005)
39. Triggs, B.: Factorization methods for projective structure and motion. In: Proc. of the IEEE Conference on Computer Vision and Pattern Recognition, pp. 845–851 (1996)
40. Vidal, R., Abretske, D.: Nonrigid shape and motion from multiple perspective views. In: Proc. of European Conference on Computer Vision. Lecture Notes in Computer Science, vol. 3952, pp. 205–218. Springer, Berlin (2006)

41. Wang, G., Tsui, H.T., Wu, J.: Rotation constrained power factorization for structure from motion of nonrigid objects. Pattern Recogn. Lett. **29**(1), 72–80 (2008)
42. Wang, G., Wu, J.: Quasi-perspective projection model: Theory and application to structure and motion factorization from uncalibrated image sequences. Int. J. Comput. Vis. **87**(3), 213–234 (2010)
43. Xiao, J., Chai, J., Kanade, T.: A closed-form solution to non-rigid shape and motion recovery. Int. J. Comput. Vis. **67**(2), 233–246 (2006)
44. Xiao, J., Chai, J.X., Kanade, T.: A closed-form solution to non-rigid shape and motion recovery. In: Proc. of European Conference on Computer Vision, vol. 4, pp. 573–587 (2004)
45. Xiao, J., Kanade, T.: Uncalibrated perspective reconstruction of deformable structures. In: Proc. of the International Conference on Computer Vision, vol. 2, pp. 1075–1082 (2005)
46. Yan, J., Pollefeys, M.: A factorization-based approach to articulated motion recovery. In: Proc. of IEEE Conference on Computer Vision and Pattern Recognition, vol. 2, pp. 815–821 (2005)
47. Yan, J., Pollefeys, M.: A factorization-based approach for articulated nonrigid shape, motion and kinematic chain recovery from video. IEEE Trans. Pattern Anal. Mach. Intell. **30**(5), 865–877 (2008)
48. Zaharescu, A., Horaud, R.: Robust factorization methods using a Gaussian/uniform mixture model. Int. J. Comput. Vis. **81**(3), 240–258 (2009)
49. Zelnik-Manor, L., Irani, M.: Degeneracies, dependencies and their implications in multi-body and multi-sequence factorizations. In: Proc. of the IEEE Conference on Computer Vision and Pattern Recognition, vol. 2, p. 287 (2003)
50. Zelnik-Manor, L., Irani, M.: On single-sequence and multi-sequence factorizations. Int. J. Comput. Vis. **67**(3), 313–326 (2006)
51. Zelnik-Manor, L., Machline, M., Irani, M.: Multi-body factorization with uncertainty: Revisiting motion consistency. Int. J. Comput. Vis. **68**(1), 27–41 (2006)
52. Zhang, Z., Deriche, R., Faugeras, O., Luong, Q.T.: A robust technique for matching two uncalibrated images through the recovery of the unknown epipolar geometry. Artif. Intell. **78**(1–2), 87–119 (1995)

Perspective 3D Reconstruction of Rigid Objects

5

Abstract It is well known that projective depth recovery and camera calibration are two essential and difficult steps in the problem of 3D Euclidean structure and motion recovery from video sequences. This chapter presents two new algorithms to improve the performance of perspective factorization. The first one is a hybrid method for projective depths estimation. It initializes the depth scales via a projective structure reconstructed from two views with large camera movement, which are then optimized iteratively by minimizing reprojection residues. The algorithm is more accurate than previous methods and converges quickly. The second one is on camera self-calibration based on Kruppa constraints which can deal with a more general camera model. Then the Euclidean structure is recovered from factorization of the normalized tracking matrix. Extensive experiments on synthetic data and real sequences are performed for validation and comparison.

Nature is an infinite sphere of which the center is everywhere and the circumference nowhere.

Blaise Pascal (1623–1662)

5.1 Introduction

Most of the studies on structure and motion factorization assume affine camera model due to its simplicity [5, 16, 19, 26]. This is a zero order (weak-perspective) or first order (para-perspective) approximation of a general perspective projection. Christy and Horaud [1] extended the method to perspective camera model by incrementally performing the affine factorization of a scaled tracking matrix. The method is an affine approximation to a general perspective projection. Triggs [21] and Sturm and Triggs [20] proposed a full projective reconstruction method via rank-4 factorization of a scaled tracking matrix with projective depths recovered from pairwise epipolar geometry. The method was further studied in [3, 11, 22], where different iterative methods were proposed to recover projective depths. Zaharescu *et al.* [28] proposed an incremental perspective factorization method. Hung and

G. Wang, Q.M.J. Wu, *Guide to Three Dimensional Structure and Motion Factorization*, Advances in Pattern Recognition,
DOI 10.1007/978-0-85729-046-5_5,

Tang [9] proposed to solve the perspective structure by integrating initial search and bundle adjustment into a set of weighted least-squares problems.

In case of uncalibrated image sequences, camera calibration is an indispensable step in retrieving 3D metric information. The theory of camera self-calibration was first introduced in computer vision by Faugeras *et al.* [2] and Maybank and Faugeras [13]. The method is based on the so-called Kruppa equation that links the epipolar transformation to the dual of the imaged absolute conic (DIAC). Camera self-calibration can be accomplished from only the image measurements without requiring special calibration objects or known camera motions. Following this idea, many improvements and related methods were proposed [4, 10, 27]. Pollefeys *et al.* [18] proposed a modulus constraint on the position of the plane at infinity for camera self-calibration under constant parameters.

Many calibration methods assume constant camera model with fixed intrinsic parameters. In order to calibrate cameras with varying parameters, Pollefeys *et al.* [17] extended the modulus constraint to retrieve metric structure from image sequences obtained with uncalibrated zooming cameras. Heyden and Åström [6] proved that self-calibration is possible when the aspect ratio is known and no skew is present. They further extended the method and generalized it by proving that the existence of one constant intrinsic parameter suffices for self-calibration [7]. However, the Kruppa equations are quadratic in nature and good initial values are required.

The general process of structure and motion recovery from uncalibrated video sequences is illustrated in Fig. 5.1. To obtain Euclidean structure from perspective factorization, there are two essential steps. One is to recover a set of consistent perspective depths; the other is to recover the camera parameters so as to upgrade the reconstruction from perspective to Euclidean. In this chapter, we try to improve the performance of existing methods with respect to these two important aspects. For projective depths, we propose a new initialization method for the optimization scheme. The initial values are very close to the ground truths, so that the solution can quickly converge to a global minimum. For camera calibration, we propose to combine the Kruppa constraints with camera constraints to calibrate a more general camera model with varying parameters. Previous studies usually assume a simplified camera model which may be invalid and cause large reconstruction errors in some cases. Experiments show good improvements over existing methods.

The remaining part of this chapter is organized as follows. Some previous studies on projective depth recovery are briefly reviewed in Sect. 5.2. The proposed hybrid method is elaborated in Sect. 5.3. The calibration and reconstruction methods are presented in Sect. 5.4. Experimental evaluations on synthetic and real images are presented in Sects. 5.5 and 5.6 respectively.

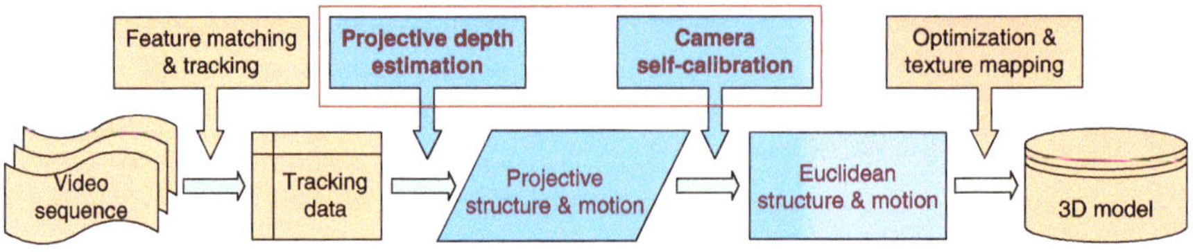

Fig. 5.1 Process of perspective Euclidean structure and motion recovery from uncalibrated video sequences

5.2 Previous Works on Projective Depths Recovery

A general form of perspective factorization of rigid objects can be written as

$$\underbrace{\begin{bmatrix} \lambda_{11}\mathbf{x}_{11} & \cdots & \lambda_{1n}\mathbf{x}_{1n} \\ \vdots & \ddots & \vdots \\ \lambda_{m1}\mathbf{x}_{m1} & \cdots & \lambda_{mn}\mathbf{x}_{mn} \end{bmatrix}}_{\mathbf{W}_{3m\times n}} = \underbrace{\begin{bmatrix} \mathbf{P}_1 \\ \vdots \\ \mathbf{P}_m \end{bmatrix}}_{\mathbf{M}_{3m\times 4}} \underbrace{[\mathbf{X}_1, \ldots, \mathbf{X}_n]}_{\mathbf{S}_{4\times n}} \tag{5.1}$$

There are two essential steps for the algorithm: the first step is to recover a set of consistent projective depths that correctly rescale the tracking matrix $\mathbf{W}$; and the second is to recover a transformation matrix to upgrade the structure from projective to the Euclidean space.

Researchers have proposed several schemes to recover the projective depths, such as the direct computation method via epipolar geometry [20, 21], iteration algorithm [3, 22], and hybrid method [25]. It is easy to verify that the projective depths for features in the same image have fixed ratios, although the depth scale for a single point is arbitrary. Similarly, the projective depths for a single point across the image sequence also have fixed ratios. Therefore, the scale factors in the tracking matrix are not unique. We will give a brief discussion of these methods below.

5.2.1 Epipolar Geometry Based Algorithm

Sturm and Triggs [20, 21] proposed a non-iterative method to recover the projective depths based on epipolar constraints. Under perspective projection, the projection of an image point j in any two frames i and i' can be formulated as

$$\begin{bmatrix} \mathbf{P}_i & | \; \lambda_{ij}\mathbf{x}_{ij} \\ \mathbf{P}_{i'} & | \; \lambda_{i'j}\mathbf{x}_{i'j} \end{bmatrix} = \begin{bmatrix} \mathbf{P}_i \\ \mathbf{P}_{i'} \end{bmatrix} [\mathbf{I}_4 | \mathbf{X}_j] \tag{5.2}$$

where $\mathbf{I}_4$ is a 4×4 identity matrix. The left part of above equation is a 6×5 matrix whose rank is no more than 4. Thus all of its 5×5 minors will vanish. If we expand these by cofactors in the last column, we will obtain a homogeneous linear equation of $\lambda_{ij}\mathbf{x}_{ij}$ and $\lambda_{i'j}\mathbf{x}_{i'j}$ with 4×4 determinants of the projection matrix rows as coefficients. These coefficients satisfy the following epipolar relationship between the epipolar lines.

$$[\mathbf{F}_{ii'}\mathbf{x}_{i'j}]\lambda_{i'j} = [\mathbf{e}_{i'i} \times \mathbf{x}_{ij}]\lambda_{ij} \tag{5.3}$$

where $\mathbf{F}_{ii'}$ is the fundamental matrix between the two frames, $\mathbf{e}_{i'i}$ is the epipole, '$\times$' stands for cross product of two vectors. Equation (5.3) establishes the relationship between projection depths of one space point in two images. Thus we can directly recover a consistent set of projective depths of a given point across the sequence, which is defined up to a scale. From (5.3), we can solve depth λ_{ij} in terms of $\lambda_{i'j}$ via least squares.

$$\lambda_{ij} = \frac{[\mathbf{e}_{i'i} \times \mathbf{x}_{ij}]^T [\mathbf{F}_{ii'} \mathbf{x}_{i'j}]}{\|\mathbf{e}_{i'i} \times \mathbf{x}_{ij}\|^2} \lambda_{i'j} \tag{5.4}$$

Simply one may select adjacent pairwise images in the sequence and calculate the projective depths of point j by joining together a single connected graph, which starts from any arbitrary initial value and one may set $\lambda_{1j} = 1$. Using this relationship, all of the projective depths can be recovered modulo row and column scalings. This strategy, however, depends highly on the accurate estimation of the fundamental matrices, which might be unstable in case of large image noise or when the tracked features are located in critical configurations [12]. Also the computation error may propagate along the computational chain and degrade the accuracy of the tracking matrix.

5.2.2 Iteration Based Algorithm

One inherent advantage of the factorization algorithm lies in its uniform treatment of all tracking data from a sequence. Taking advantage of this property, Ueshiba and Tomita [22] proposed an iterative nonlinear method to update the projective depths so as to satisfy the rank 4 constraint of the tracking matrix.

By performing SVD decomposition on the tracking matrix, we can obtain the following expression.

$$\begin{aligned} \mathbf{W} &= \mathbf{U}\,\mathrm{diag}(\sigma_1, \sigma_2, \ldots, \sigma_N)\mathbf{V}^T \\ &= \underbrace{\mathbf{U}_1\,\mathrm{diag}(\sigma_1, \ldots, \sigma_4)\mathbf{V}_1^T}_{\tilde{\mathbf{W}}} + \underbrace{\mathbf{U}_2\,\mathrm{diag}(\sigma_5, \ldots, \sigma_N)\mathbf{V}_2^T}_{\Delta\mathbf{W}} \end{aligned} \tag{5.5}$$

where $\sigma_1 \geq \sigma_2 \geq \cdots \geq \sigma_N \geq 0$ are singular values of the tracking matrix with $N = \min(3m, n)$, and $\mathbf{U}$ and $\mathbf{V}$ are orthogonal matrices, $\mathbf{U}_1$ and $\mathbf{U}_2$ represent the first 4 and the remaining columns of $\mathbf{U}$, similarly, $\mathbf{V}_1$ and $\mathbf{V}_2$ represent the first 4 and the remaining rows of $\mathbf{V}$. Assuming the tracking matrix is noise free and a set of consistent projective depths are correctly recovered, $\mathbf{W}$ is exactly rank 4, thus $\tilde{\mathbf{W}} = \mathbf{W}$ and $\Delta\mathbf{W} = \mathbf{0}$.

When the projective depths are unknown, the rank of the tracking matrix will be greater than 4. The following cost function can be employed to enforce the rank constraint.

$$J_\lambda = \min_{\lambda_{ij}} \|\Delta\mathbf{W}\|^2 = \min_{\lambda_{ij}} \sum_{i=5}^{N} \sigma_i^2 \tag{5.6}$$

The projective depths are estimated by minimizing summation of the remaining singular values. The minimization process can be performed via a conjugate gradient descent algorithm, which starts from affine assumption as $\lambda_{ij} = 1$. Please refer to [22] for computation details.

Alternatively, Han and Kanade [3] proposed a simpler method, which also starts from affine. They iteratively perform a rank-4 factorization on the weighted tracking matrix to generate the projective motion and shape matrices, and update the depths via $\lambda_{ij} = \mathbf{P}_i^3 \mathbf{X}_j$ until convergence, where $\mathbf{P}_i^3$ denotes the third row of $\mathbf{P}_i$. They also discussed the possibility to calibrate multiple camera parameters under the factorization framework.

However, there is no guarantee that the above two iteration procedures will converge to a global minimum. As proved more recently by Oliensis and Hartley [14], no iteration has been shown to converge sensibly. The assumption of $\lambda_{ij} = 1$ is only valid when the ratios of true depths of different 3D points remain approximately constant throughout the sequence. For instance an aerial image sequence captured by an onboard camera pointing straight down from constant altitude.

5.3 Hybrid Projective Depths Recovery

In this section, we will introduce a hybrid method to recover projective depths via an optimization scheme, which is initialized using a projective structure obtained from a pair of reference frames.

5.3.1 Initialization and Optimization

Suppose the fundamental matrix of the first two frames is $\mathbf{F}_{12}$. Then the canonical projection matrices of the two frames can be expressed as follows [5].

$$\begin{cases} \mathbf{P}_1 = [\mathbf{I} \,|\, \mathbf{0}] \\ \mathbf{P}_2 = \left[[\mathbf{e}_{21}]_\times \mathbf{F}_{12} + \mathbf{e}_{21}\mathbf{v}^T \,|\, \alpha \mathbf{e}_{21} \right] \end{cases} \tag{5.7}$$

where $\mathbf{e}_{21}$ is the epipole in the second view, $[\mathbf{e}_{21}]_\times$ stands for the 3×3 antisymmetric matrix derived from vector $\mathbf{e}_{21}$, $\mathbf{v}$ is an arbitrary 3-vector, and α is a nonzero scalar. For any space point $\mathbf{X}_j$, its image in the two views is given by

$$\begin{cases} \lambda_{1j}\mathbf{x}_{1j} = \mathbf{P}_1 \mathbf{X}_j \\ \lambda_{2j}\mathbf{x}_{2j} = \mathbf{P}_2 \mathbf{X}_j \end{cases} \tag{5.8}$$

The equations can be written as

$$\mathbf{A}_j \mathbf{X}_j = \begin{bmatrix} [\mathbf{x}_{1j}]_\times \mathbf{P}_1 \\ [\mathbf{x}_{2j}]_\times \mathbf{P}_2 \end{bmatrix} \mathbf{X}_j = \mathbf{0} \tag{5.9}$$

where $\mathbf{A}_j$ is a 6×4 matrix of rank 3, since the two features satisfy the epipolar constraint. Thus we can obtain an initial estimate of the perspective structure $\mathbf{X}_j$ from the right null space of $\mathbf{A}_j$. Then from perspective projection we have the following constraints on the

5

projection matrix of any frame i.

$$\begin{bmatrix} \mathbf{X}_j^T & \mathbf{0}^T & -u_{ij}\mathbf{X}_j^T \\ \mathbf{0}^T & \mathbf{X}_j^T & -v_{ij}\mathbf{X}_j^T \end{bmatrix} \begin{bmatrix} \mathbf{P}_i^{1T} \\ \mathbf{P}_i^{2T} \\ \mathbf{P}_i^{3T} \end{bmatrix} = \mathbf{0} \tag{5.10}$$

where $\mathbf{P}_i^k$ denotes the k-th row of $\mathbf{P}_i$. Each point provides two constraints, and the projection matrix can be computed linearly from $n \geq 6$ points in a general position. After retrieving all the projection matrices, the projection depths can be initialized from perspective projection equation via least squares.

$$J_1 = \frac{1}{2} \min_{\hat{\lambda}_{ij}} \|\hat{\lambda}_{ij}\mathbf{x}_{ij} - \mathbf{P}_i\mathbf{X}_j\|_F^2, \quad i = 1, \ldots, m, j = 1, \ldots, n \tag{5.11}$$

where $\|\cdot\|_F$ stands for the Frobenius norm of a vector. Ideally, $\hat{\lambda}_{ij}$ should correspond to a set of correct projective depths. However, this is not true due to noise contamination. The solution is used as an initial value and the projective depths are iteratively optimized until convergence. The optimization is achieved as follows.

1. Update current tracking matrix $\mathbf{W}_t$ with the estimated depth scales and perform a rank-4 factorization on the weighted tracking matrix to generate a set of projective motion $\mathbf{M}_t$ and shape $\mathbf{S}_t$ matrices.
2. Update the projective depths by minimizing the reprojection residuals as

$$J_2 = \frac{1}{2} \min_{\lambda_{ij}} \|\mathbf{W}_t - \mathbf{M}_t\mathbf{S}_t\|_F^2 \tag{5.12}$$

The minimization problem is solved via least-squares. During iterations, the convergence may be determined by checking the variation of the projective depths with respect to the previous iteration.

$$\varepsilon = \frac{1}{mn} \sum_{i=1}^{m} \sum_{j=1}^{n} |\Delta\lambda_{ij}|^2 \tag{5.13}$$

The optimization process is similar to [3], however, there are two major differences. First, the projective depths in [3] were updated from $\lambda_{ij} = \mathbf{P}_i^3\mathbf{X}_j$. This is not optimal for noise contaminated data, since the corresponding inhomogeneous point $[\mathbf{P}_i^1\mathbf{X}_j/\lambda_{ij}, \mathbf{P}_i^2\mathbf{X}_j/\lambda_{ij}]^T$ usually does not correspond to a true image point. Second, the iteration in [3] was started from $\lambda_{ij} = 1$, and it may not converge to a global minimum. The proposed initialization is very close to the true solution, thus the iteration usually converges quickly to the global minimum.

Remark 5.1 It should be noted that the recovered projective depths may not be unique. Suppose a depth λ_{ij} satisfies $\lambda_{ij}\mathbf{x}_{ij} = \mathbf{P}_i\mathbf{X}_j$, then if we replace $\mathbf{P}_i$ with $\mu_i\mathbf{P}_i$, and $\mathbf{X}_j$ with $\ell_j\mathbf{X}_j$, we have

$$(\mu_i\ell_j\lambda_{ij})\mathbf{x}_{ij} = (\mu_i\mathbf{P}_i)(\ell_j\mathbf{X}_j) \tag{5.14}$$

Therefore, each triple-row of the tracking matrix in (5.1) may be multiplied by a scalar μ_i, which is equivalent to multiplying $\mathbf{P}_i$ with μ_i. Similarly, each column of the tracking matrix may be multiplied by a scalar ℓ_j, which is equivalent to multiplying $\mathbf{X}_j$ with μ_j.

Let us reconsider the minimization criteria in (5.12). Suppose the projective depths and the weighted tracking matrix are λ_{ij} and $\mathbf{W}$ at current iteration, and $\hat{\lambda}_{ij}$ and $\hat{\mathbf{W}}$ at next iteration. Then the minimization can be expressed as

$$\begin{aligned} J_2' &= \frac{1}{2}\min \|\mathbf{W} - \hat{\mathbf{W}}\|_F^2 \\ &= \frac{1}{2}\min \sum_{i=1}^{m}\sum_{j=1}^{n} \|\lambda_{ij}\mathbf{x}_{ij} - \hat{\lambda}_{ij}\hat{\mathbf{x}}_{ij}\|_F^2 \\ &= \frac{1}{2}\min \sum_{i=1}^{m}\sum_{j=1}^{n} \left((\lambda_{ij}u_{ij} - \hat{\lambda}_{ij}\hat{u}_{ij})^2 + (\lambda_{ij}v_{ij} - \hat{\lambda}_{ij}\hat{v}_{ij})^2 + (\lambda_{ij} - \hat{\lambda}_{ij})^2\right) \end{aligned} \tag{5.15}$$

It is clear that the terms to be minimized in (5.15) do not have any geometric meaning, and the estimated depths tend to be close to λ_{ij} due to the effect of the third term. If all projective depths are very close to each other, the cost function tends to minimize $\sum_{ij}\|\mathbf{x}_{ij} - \hat{\mathbf{x}}_{ij}\|_F^2$, which is the geometric distance between the estimated image points.

Remark 5.2 After recovering the projective depths, we can perform factorization directly on the weighted tracking matrix $\mathbf{W} = \{\lambda_{ij}\mathbf{x}_{ij} | i = 1, \ldots, m, j = 1, \ldots, n\}$. However, this is not an optimal solution as the tracking matrix may be poorly conditioned. Since the projective depths are not unique as noted in Remark 5.1, we adopt a similar method as in [20] to normalize the tracking matrix so as to increase its numerical stability and accuracy. First, we rescale $\mathbf{W}$ image-wisely by multiplying every triple-row by a scalar as $\mu_i\mathbf{x}_{ij} \to \mathbf{x}_{ij}$ such that they have unit norm. Similarly, a normalization is implemented in a point-wise manner by multiplying a scalar as $\ell_j\mathbf{x}_{ij} \to \mathbf{x}_{ij}$ to each column of $\mathbf{W}$. The process can be iterated several times.

5.3.2 Selection of Reference Frames

In the above analysis, we utilized the first two frames to estimate the projective structure. The results may be unreliable for small or degenerate inter-view motion. To achieve a more robust and reliable reconstruction, it is better to choose two views with large rotation and translation amongst them. We adopt the following criteria to choose the images.

Let us take the rth frame as a reference, and define a disparity matrix $\mathbf{D}$ as

$$\mathbf{D} = \begin{bmatrix} \|\mathbf{x}_{11} - \mathbf{x}_{r1}\|_F & \cdots & \|\mathbf{x}_{1n} - \mathbf{x}_{rn}\|_F \\ \vdots & \ddots & \vdots \\ \|\mathbf{x}_{m1} - \mathbf{x}_{r1}\|_F & \cdots & \|\mathbf{x}_{mn} - \mathbf{x}_{rn}\|_F \end{bmatrix} \tag{5.16}$$

5

where $\|\mathbf{x}_{ij} - \mathbf{x}_{rj}\|_F$ is the disparity between the two points, the ith row of $\mathbf{D}$ stands for the point disparities of the ith frame to the reference frame. Let us calculate the mean and standard deviation of each row and denote them as

$$\mathbf{D}_m = \begin{bmatrix} e_1 \\ \vdots \\ e_m \end{bmatrix}, \quad \mathbf{D}_s = \begin{bmatrix} s_1 \\ \vdots \\ s_m \end{bmatrix} \tag{5.17}$$

Generally speaking, the values of rotation and translation with respect to the reference frame can be reflected by its mean $\mathbf{D}_m$ and standard deviation $\mathbf{D}_s$. Let us define

$$[d_{max}, r'] = \max\left(\frac{\mathbf{D}_m \odot \mathbf{D}_s}{\|\mathbf{D}_m\|_F \|\mathbf{D}_s\|_F}\right) \tag{5.18}$$

where '$\odot$' denotes element-by-element multiplication, d_{max} is the largest value and r' is its index in the vector. Then the frame r' will exhibit a large camera movement with respect to the reference frame r. Thus we may take the two frames as reference to estimate the initial perspective structure.

5.4 Camera Calibration and Euclidean Reconstruction

In this section, we propose a two-step scheme to calibrate the camera. The focal lengths are initially estimated from perspective structure and motion factorization of the tracking matrix, then all camera parameters are further refined via Kruppa constraints.

5.4.1 Camera Self-calibration

After recovering a set of correct projective depths, the weighted tracking matrix is of rank 4, and it can be easily factorized into a motion matrix $\mathbf{M}$ and shape matrix $\mathbf{S}$. However, the decomposition gives only one solution in projective space which is not unique. Thus we have to find a nonsingular linear transformation $\mathbf{H} \in \mathbb{R}^{4\times4}$ to upgrade the solution to the Euclidean space.

For the recovery of transformation matrix $\mathbf{H}$, many researchers [3, 19, 24] have adopted a metric constraint on the motion matrix, which is indeed a self-calibration process derived from the dual image of the absolute conic (DIAC) $\mathbf{C}_i = \mathbf{K}_i\mathbf{K}_i^T$, where $\mathbf{K}_i$ is the camera calibration matrix of the form

$$\mathbf{K}_i = \begin{bmatrix} f_i & \varsigma_i & u_{0i} \\ 0 & \kappa_i f_i & v_{0i} \\ 0 & 0 & 1 \end{bmatrix} \tag{5.19}$$

In order to explore the metric constraint, we assume a simplified camera model with zero skew, unit aspect ratio and known principal point, i.e. $\varsigma_i = 0, \kappa_i = 1$ and $u_{0i} = v_{0i} = 0$. Then the camera model is simplified and contains only one intrinsic parameter f_i. Nevertheless, the assumption is not usually satisfied for many digital cameras. We will now calibrate the camera parameters in a more general case via Kruppa constraint.

Given two frames i and i', and the fundamental matrix $\mathbf{F}_{ii'}$ between the images. Let us perform SVD decomposition on $\mathbf{F}_{ii'}$ as

$$\mathbf{F}_{ii'} = \mathbf{U}\Sigma\mathbf{V}^T = [\mathbf{u}_1, \mathbf{u}_2, \mathbf{u}_3] \begin{bmatrix} \sigma_1 & & \\ & \sigma_2 & \\ & & 0 \end{bmatrix} [\mathbf{v}_1, \mathbf{v}_2, \mathbf{v}_3]^T \tag{5.20}$$

where the subscript notation $'i'$ is omitted for simplicity. σ_1 and σ_2 are the two singular values of $\mathbf{F}_{ii'}$ since the fundamental matrix is of rank 2. Then the following Kruppa equation can be obtained from epipolar geometry [5].

$$\eta_i \begin{bmatrix} \mathbf{u}_2^T \mathbf{C}_{i'} \mathbf{u}_2 \\ -\mathbf{u}_1^T \mathbf{C}_{i'} \mathbf{u}_2 \\ \mathbf{u}_1^T \mathbf{C}_{i'} \mathbf{u}_1 \end{bmatrix} = \begin{bmatrix} \sigma_1^2 \mathbf{v}_1^T \mathbf{C}_i \mathbf{v}_1 \\ \sigma_1 \sigma_2 \mathbf{v}_1^T \mathbf{C}_i \mathbf{v}_2 \\ \sigma_2^2 \mathbf{v}_2^T \mathbf{C}_i \mathbf{v}_2 \end{bmatrix} \tag{5.21}$$

where

$$\mathbf{C}_i = \mathbf{K}_i \mathbf{K}_i^T = \begin{bmatrix} f_i^2 + \varsigma_i^2 + u_{0i}^2 & \varsigma_i \kappa_i f_i + u_{0i} v_{0i} & u_{0i} \\ \varsigma_i \kappa_i f_i + u_{0i} v_{0i} & \kappa_i^2 f_i^2 + v_{0i}^2 & v_{0i} \\ u_{0i} & v_{0i} & 1 \end{bmatrix} \tag{5.22}$$

and $\mathbf{C}_{i'} = \mathbf{K}_{i'} \mathbf{K}_{i'}^T$. Thus we can obtain three quadratic constraints by eliminating the unknown scalar η_i.

$$\begin{cases} g_{k1} = -\sigma_2^2 \mathbf{u}_1^T \mathbf{C}_{i'} \mathbf{u}_2 \mathbf{v}_2^T \mathbf{C}_i \mathbf{v}_2 - \sigma_1 \sigma_2 \mathbf{u}_1^T \mathbf{C}_{i'} \mathbf{u}_1 \mathbf{v}_1^T \mathbf{C}_i \mathbf{v}_2 = 0 \\ g_{k2} = \sigma_1^2 \mathbf{u}_1^T \mathbf{C}_{i'} \mathbf{u}_1 \mathbf{v}_1^T \mathbf{C}_i \mathbf{v}_1 - \sigma_2^2 \mathbf{u}_2^T \mathbf{C}_{i'} \mathbf{u}_2 \mathbf{v}_2^T \mathbf{C}_i \mathbf{v}_2 = 0 \\ g_{k3} = \sigma_1 \sigma_2 \mathbf{u}_2^T \mathbf{C}_{i'} \mathbf{u}_2 \mathbf{v}_1^T \mathbf{C}_i \mathbf{v}_2 + \sigma_1^2 \mathbf{u}_1^T \mathbf{C}_{i'} \mathbf{u}_2 \mathbf{v}_1^T \mathbf{C}_i \mathbf{v}_1 = 0 \end{cases} \tag{5.23}$$

where g_{k1}, g_{k2}, and g_{k3} are the functions defined in terms of the 10 elements of $\mathbf{C}_i$ and $\mathbf{C}_{i'}$, but only two of them are independent since (5.21) is defined up to a scale. Given $m(\geq 3)$ images, we have $\frac{1}{2}m(m-1)$ fundamental matrices and $\frac{3}{2}m(m-1)$ Kruppa constraints (5.23). However, as proved in [8], only $(5m-9)$ of the equations are independent. If we have no prior information about the cameras, there will be $5m$ intrinsic parameters, it is impossible to calibrate the cameras. For an image sequence captured by one camera, usually it is safe to assume that the principal point $[u_{0i}, v_{0i}]^T$ is fixed during shooting. Thus we add the following two constraints to $\mathbf{C}_i$.

$$\begin{cases} g_{c1} = \mathbf{C}_i(1,3) - \mathbf{C}_{i'}(1,3) = 0 \\ g_{c2} = \mathbf{C}_i(2,3) - \mathbf{C}_{i'}(2,3) = 0 \end{cases} \tag{5.24}$$

Given m images, there will be $(3m+2)$ intrinsic camera parameters, and 6 images are sufficient to calibrate the cameras. If we further assume zero skew (i.e. $\varsigma_i = 0$), then we will have one additional constraint as

$$g_{c3} = \mathbf{C}_i(1,2) - \mathbf{C}_{i'}(1,2) = 0$$

and the unknowns are reduced to $(2m+3)$. Thus the cameras can be calibrated from 4 images. Under the constraints (5.23) to (5.24), we can formulate the calibration problem as the following minimization scheme.

$$J_3 = \frac{1}{2} \min_{\mathbf{C}_i, \mathbf{C}_{i'}} \sum_{i=1}^{m} \sum_{i'=i+1}^{m} \left(\tau (g_{k1}^2 + g_{k2}^2 + g_{k3}^2) + (1-\tau)(g_{c1}^2 + g_{c2}^2) \right) \tag{5.25}$$

where τ is a weighting coefficient. The minimization process can be easily solved via gradient descent algorithm. Then the camera parameters are factorized from $\mathbf{C}_i$ via Cholesky decomposition.

5.4.2 Euclidean Reconstruction

Using the recovered camera parameters, we can normalize all images by the camera parameters, i.e. apply the inverse of camera matrix to the corresponding image as $\tilde{\mathbf{x}}_{ij} \leftarrow \mathbf{K}_i^{-1}\mathbf{x}_{ij}$. Then the factorization of (5.1) is modified to

$$\begin{bmatrix} \lambda_{11}\tilde{\mathbf{x}}_{11} & \cdots & \lambda_{1n}\tilde{\mathbf{x}}_{1n} \\ \vdots & \ddots & \vdots \\ \lambda_{m1}\tilde{\mathbf{x}}_{m1} & \cdots & \lambda_{mn}\tilde{\mathbf{x}}_{mn} \end{bmatrix} = \begin{bmatrix} \mu_1\mathbf{R}_1 & \mu_1\mathbf{T}_1 \\ \vdots & \vdots \\ \mu_m\mathbf{R}_m & \mu_m\mathbf{T}_m \end{bmatrix} [\ell_1\mathbf{X}_1, \ldots, \ell_n\mathbf{X}_n] \tag{5.26}$$

Let us write it in short as $\tilde{\mathbf{W}}_{3m\times n} = \tilde{\mathbf{M}}_{3m\times 4}\mathbf{S}_{4\times n}$. By performing SVD on the normalized tracking matrix and imposing the rank constraint, $\tilde{\mathbf{W}}$ may be factored as $\hat{\mathbf{M}}_{3m\times 4}\hat{\mathbf{S}}_{4\times n}$. The decomposition is only defined up to a non-singular 4×4 linear transformation $\mathbf{H}_{4\times 4}$ as $\tilde{\mathbf{M}} = \hat{\mathbf{M}}\mathbf{H}$ and $\mathbf{S} = \mathbf{H}^{-1}\hat{\mathbf{S}}$.

Similar to the study in [24], we utilize the metric information to compute the transformation matrix. Let us denote $\mathbf{H}_{4\times 4} = [\mathbf{H}_l | \mathbf{H}_r]$, where $\mathbf{H}_l$ corresponds to the first three columns of $\mathbf{H}$ and $\mathbf{H}_r$ represents the last column of $\mathbf{H}$. Suppose $\hat{\mathbf{M}}_i$ is the ith triple-row of $\hat{\mathbf{M}}$, then from $\hat{\mathbf{M}}_i\mathbf{H} = [\hat{\mathbf{M}}_i\mathbf{H}_l | \hat{\mathbf{M}}_i\mathbf{H}_r]$ we obtain

$$\begin{cases} \hat{\mathbf{M}}_i\mathbf{H}_l = \mu_i\mathbf{R}_i \\ \hat{\mathbf{M}}_i\mathbf{H}_r = \mu_i\mathbf{T}_i \end{cases} \tag{5.27}$$

Let us denote $\mathbf{N}_i = \hat{\mathbf{M}}_i \mathbf{Q} \hat{\mathbf{M}}_i^T$, where $\mathbf{Q} = \mathbf{H}_l \mathbf{H}_l^T$ is a 4×4 positive semidefinite symmetric matrix with 9 degrees of freedom. Then we have

$$\mathbf{N}_i = \hat{\mathbf{M}}_i \mathbf{Q} \hat{\mathbf{M}}_i^T = (\mu_i \mathbf{R}_i)(\mu_i \mathbf{R}_i)^T = \mu_i^2 \begin{bmatrix} 1 & 0 & 0 \\ 0 & 1 & 0 \\ 0 & 0 & 1 \end{bmatrix} \tag{5.28}$$

from which we obtain the following 5 linear constraints on the elements of $\mathbf{Q}$.

$$\begin{cases} \mathbf{N}_i(1,1) = \mathbf{N}_i(2,2) = \mathbf{N}_i(3,3) \\ \mathbf{N}_i(1,2) = \mathbf{N}_i(1,3) = \mathbf{N}_i(2,3) = 0 \end{cases} \tag{5.29}$$

Since factorization (5.26) is defined up to a global scalar, we may set $\mu_1 = 1$ to avoid the trivial solution of $\mathbf{Q} = \mathbf{0}$. Thus, we have $(5m + 1)$ linear constraints, and $\mathbf{Q}$ can be solved via least-squares. Then the matrix $\mathbf{H}_l$ is factorized from $\mathbf{Q}$, while $\mathbf{H}_r$ can be computed in the same way as in [24].

After recovering the transformation matrix $\mathbf{H}$, the Euclidean structure, rotation and translation parameters can be recovered directly from the motion matrix $\tilde{\mathbf{M}}$ and the shape matrix $\mathbf{S}$. Furthermore, the solution can be optimized via bundle adjustment by minimizing image reprojection residuals as follows.

$$J_4 = \frac{1}{2} \min_{(\mathbf{K}_i, \mathbf{R}_i, \mathbf{T}_i, \mathbf{X}_j)} \sum_{i=1}^{m} \sum_{j=1}^{n} \|\mathbf{x}_{ij} - \hat{\mathbf{x}}_{ij}\|_F^2 \tag{5.30}$$

where $\hat{\mathbf{x}}_{ij}$ denotes the reprojected image point computed from perspective projection. The minimization process is termed as bundle adjustment, which can be solved via Levenberg-Marquardt iterations [5].

Remark 5.3 According to the analysis in above sections, we need at least 7 correspondences to compute the fundamental matrix, and 4 frames to calibrate cameras. Thus the minimum requirement for the proposed scheme is 7 feature points across 4 frames in a general position. However, it is not reliable to work with minimum noise contaminated data. The proposed algorithm assumes no outliers in the tracking data, thus we utilize all available features to solve the result in least-squares sense. In presence of outliers, one should adopt a robust estimation technique, such as RANSAC [5, 15], to eliminate outliers during fundamental matrix computation.

5.4.3 Outline of the Algorithm

The implementation of the proposed method is outlined as follows.

1. Seek the correspondences of feature points across all frames;
2. Eliminate outliers via robust fundamental estimation technique [5, 15];

3. Select two reference frames with large translation and rotation according to (5.18);
4. Estimate an initial structure from the two reference frames (5.9), and initialize all projective depths according to (5.11);
5. Normalize the weighted tracking matrix $\mathbf{W}$ according to Remark 5.2;
6. Optimize projective depths iteratively by minimizing the cost function (5.12);
7. Estimate focal lengths via the methods in [24] or [3];
8. Calibrate all camera parameters according to (5.25);
9. Balance the normalized tracking matrix $\tilde{\mathbf{W}}$ as described in Remark 5.2;
10. Recover structure and motion parameters from factorization of (5.26);
11. Optimize the solution via bundle adjustment (5.30).

5.5 Evaluations on Synthetic Data

We randomly generated 100 space points within a range of $20 \times 20 \times 20$, and simulated a sequence of 10 images from these points by perspective projection. The image size is set at 800×800. Camera parameters are set as follows: Focal lengths vary randomly from 1000 to 1100, principal point is set at $[6, 8]^T$, aspect ratio $\kappa_i = 1.1$, and skew $\varsigma_i = 4$.

5.5.1 Projective Depths Recovery

We recovered the projective depths according to the proposed scheme. As a comparison, we also implemented the methods by Ueshiba [22], Han [3], and Sturm [20]. The relative errors of the recovered depths are shown in Fig. 5.2, where only the errors of the first 5 frames are displayed, and the results are obtained under 1 pixel noise level (i.e. the standard deviation of Gaussian white noise added to the image features). The histogram distributions of the errors at two different noise levels are shown in Fig. 5.3 using 7 bins. The corresponding mean and standard deviation of the errors are listed in Table 5.1. Results show that the proposed method performs better than other methods. The accuracy by Sturm's method is not good since the algorithm recovers the projective depths directly from the fundamental matrices without optimization.

We compared the average computation time and iteration times in recovering the projective depths by different algorithms. The program was implemented with Matlab R14 on Dell Inspiron 600 m laptop of Pentium(R) 1.8 GHz CPU. The average computation time for different data sets (we vary the frame number from 10 to 250) are shown in Fig. 5.4. The average iteration times are listed in Table 5.1. We see that the proposed method converges much faster than the other two iteration methods due to better initialization of projective depths. The computation time is comparable with the non-iterative method 'Sturm' and is significantly lower than the time taken by 'Ueshiba' and 'Han', especially for large data sets.

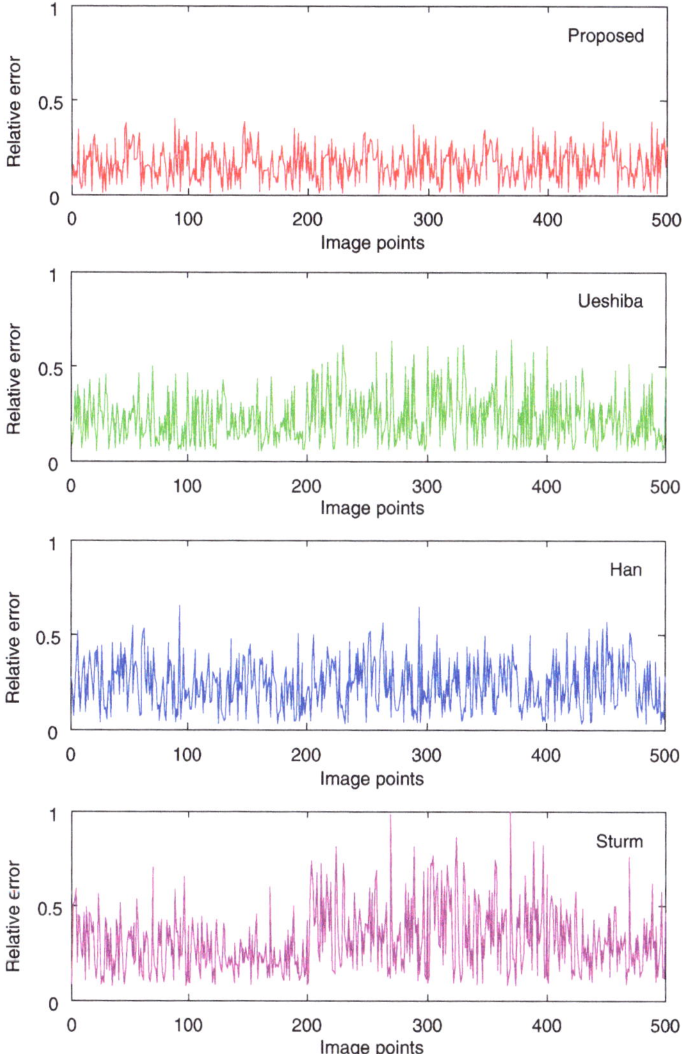

Fig. 5.2 The relative errors of all projective depths in the first 5 frames by four different methods (noise level $\sigma = 1$-pixel)

5.5.2 Calibration and Reconstruction

We calibrated cameras using the proposed algorithm, and compared it with the result obtained using the simplified camera model [24]. The relative error of the focal lengths and the recovered aspect ratio associated with each view are shown in Fig. 5.5, where the results are obtained under 1-pixel Gaussian noise. The recovered principal point and skew by

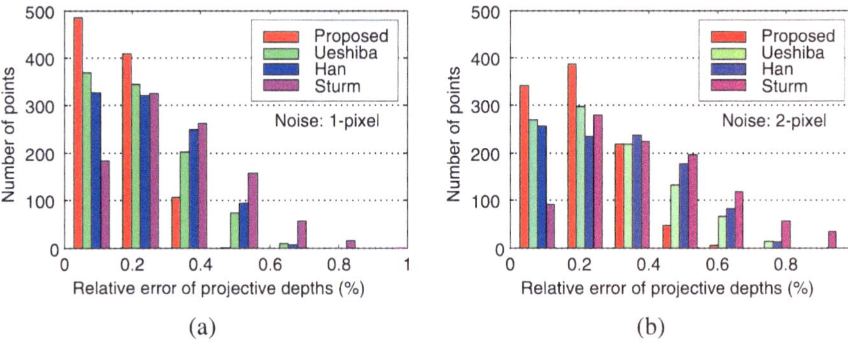

Fig. 5.3 The histogram distribution of relative errors of the recovered projective depths by four different methods. (**a**) Noise level $\sigma = 1$-pixel; (**b**) Noise level $\sigma = 2$-pixel

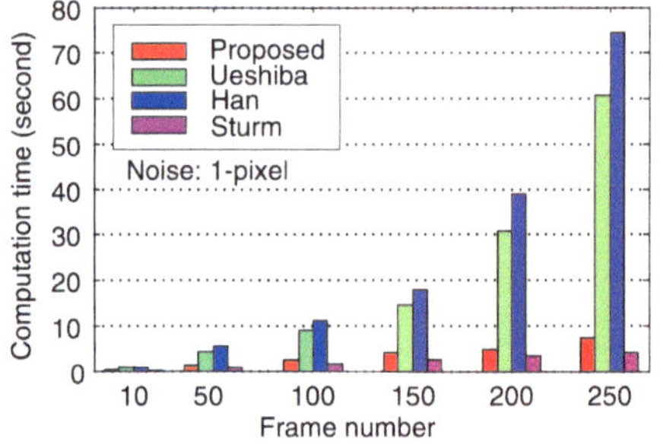

Fig. 5.4 The average computation time (in second) for different data sets by the proposed method, Ueshiba [22], Han [3], and Sturm [20]. The time is calculated with the frame number vary from 10 to 250, and noise level is set at $\sigma = 1$-pixel

Table 5.1 Relative error (%) of the recovered projective depths and iteration times at different noise levels

Method		Proposed	Ueshiba	Han	Sturm
	Mean	0.1653	0.2257	0.2416	0.3174
$\sigma = 1$-pixel	STD	0.0964	0.1329	0.1407	0.1686
	Iterations	12.5	58.4	64.9	0
	Mean	0.2091	0.2821	0.3020	0.3915
$\sigma = 2$-pixel	STD	0.1235	0.1661	0.1759	0.2007
	Iterations	14.6	59.8	67.3	0

the proposed method are $[7.25, 8.63]^T$ and 3.17, respectively. The proposed scheme can calibrate more general cameras, while the 'Simplified' algorithm only recovers the focal lengths.

We reconstructed the 3D structure of the object from the generated data, and registered the structure with the ground truth. During the test, we added Gaussian image noise to the data and varied the noise level from 0 to 3 pixels in steps of 0.5. We define reconstruction error as the pointwise distance between the recovered structure and the ground truth. The mean and standard deviation of the distances are shown in Fig. 5.6. We evaluated the reconstruction accuracy under two situations. First, we utilize the four different algorithms

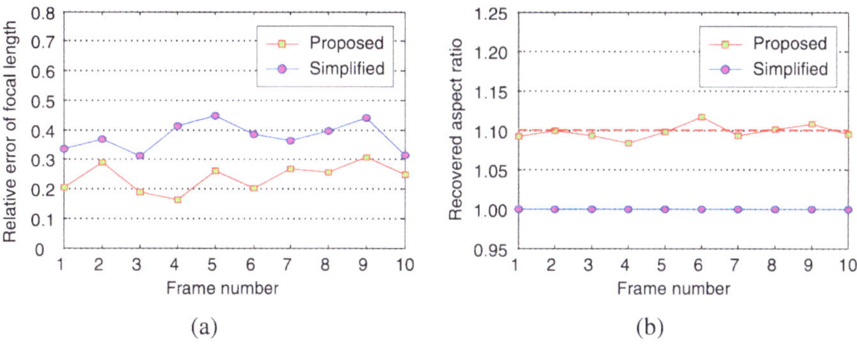

Fig. 5.5 Calibration results by the proposed method and that under simplified camera model. (**a**) The relative error of the recovered focal lengths; (**b**) The recovered value of the aspect ratio

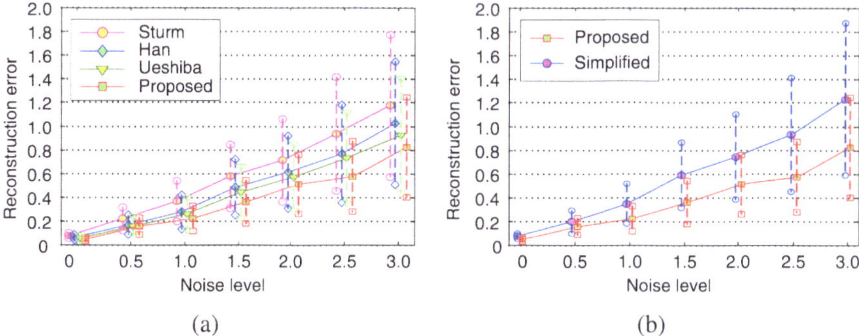

Fig. 5.6 The mean and standard deviation of the reconstruction errors under different noise levels. (**a**) Different projective depth recovery methods with the proposed calibration algorithm; (**b**) The proposed depth recovery algorithm with different calibration methods

to recover the projective depths and use the proposed method to calibrate the cameras. Second, we use the proposed algorithm to recover the depths and adopt different methods to calibrate the cameras. The proposed scheme performs much better under both cases and obtains significant improvements in depth recovery and calibration.

5.6 Evaluations on Real Sequences

The method was tested on many real image and video sequences. We will report the results on three data sets. The first two sequences in the test were captured by Canon Powershot G3 camera with a resolution of 1024×768. The last sequence on Medusa head was downloaded from Dr. Pollefeys' homepage. They were recorded by Sony TRV900 camcorder with a resolution of 720×576. All feature tracking and correspondences in these tests were established by a feature matching system based on SIFT and epipolar constraints [23].

5.6.1 Test on Model House Sequence

The model house sequence consists of 10 images taken in a gift shop. Totally 1784 reliable features were tracked across the sequence as shown in Fig. 5.7. We recover the projective depths and calibrate the cameras by the proposed scheme, then reconstruct the 3D structure of the scene. The recovered camera parameters of the first frame are listed in Table 5.2. Figure 5.7 shows the reconstructed VRML model with texture mapping and the corresponding triangulated wireframe viewed from different viewpoints. After reconstruction, we repro-

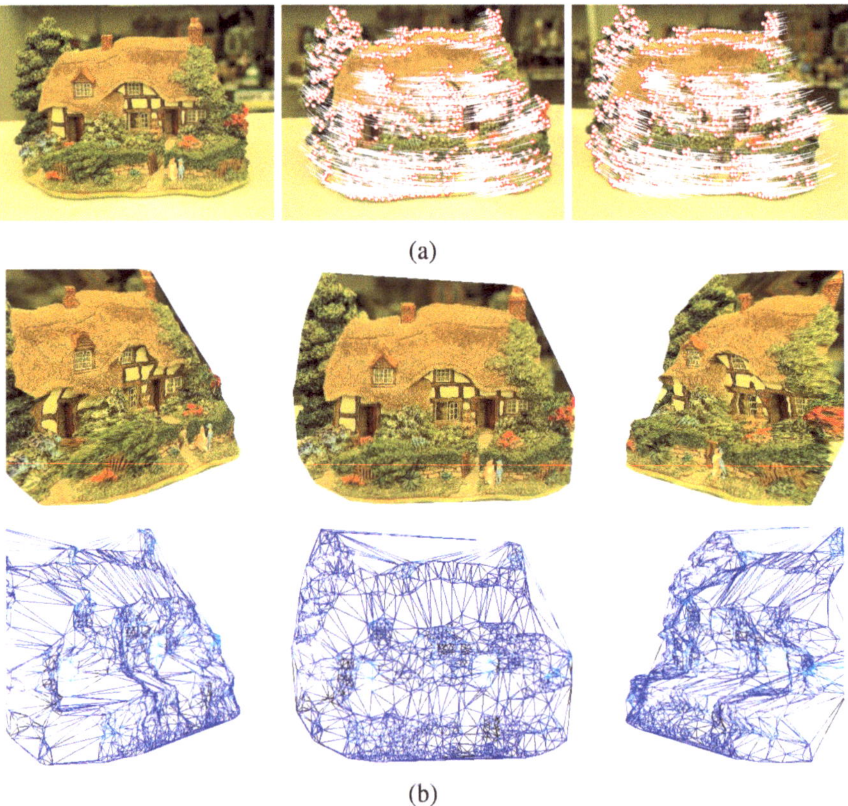

Fig. 5.7 Reconstruction results of model house sequence. (**a**) Three frames from the sequence, where the tracked features with relative disparities are overlaid to the last two images; (**b**) The reconstructed VRML model of the scene and the corresponding triangulated wireframe of the model shown from different viewpoints

Table 5.2 Calibration result of the first frame and the mean and standard deviation of reprojection errors in real sequence tests

Parameters		f_1	$\kappa_1 f_1$	u_{01}	v_{01}	ς_1	Mean	STD
House	Proposed	2890.4	2997.5	4.38	3.64	3.52	0.185	0.140
	Simplified	2823.4	2823.4	–	–	–	0.238	0.183
Post	Proposed	2101.7	2182.7	3.56	2.17	3.18	0.175	0.125
	Simplified	2154.6	2154.6	–	–	–	0.237	0.164
Head	Proposed	1315.9	1431.6	3.27	2.14	3.66	0.256	0.197
	Simplified	1376.4	1376.4	–	–	–	0.314	0.233

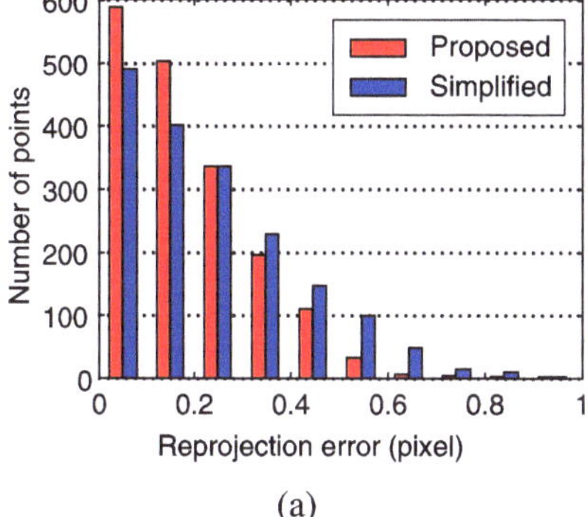

(a)

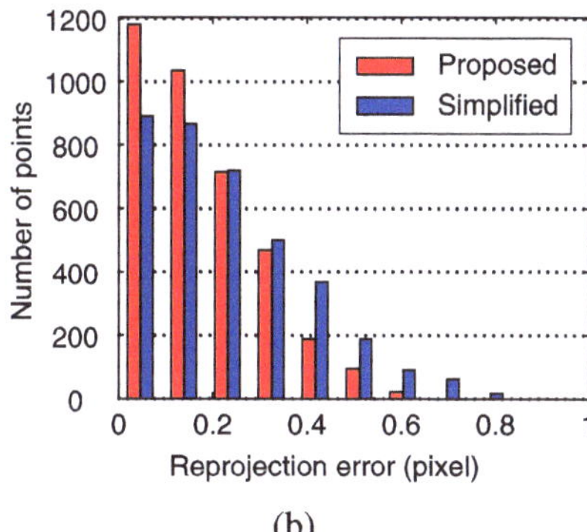

(b)

Fig. 5.8 The histogram distribution of the reprojection errors of the real sequence tests. (**a**) The model house sequence; (**b**) The stone post sequence

ject the 3D points to the images and calculate the reprojection errors. Figure 5.8 shows the histogram distributions of the errors using 10 bins. The corresponding mean and standard deviation of the errors are listed in Table 5.2. The reconstructed model is visually plausible and realistic. The reprojection error using the proposed method is smaller than that obtained using a simplified camera model.

5.6.2 Test on Stone Post Sequence

The stone post sequence consists of 8 images taken in the Sculpture Park of Windsor. We established 3693 reliable features across the sequence. Figure 5.9 shows the tracked features and the reconstructed 3D model from different viewpoints. The structure is correctly recovered using the proposed method. The histogram distributions of the reprojection errors by different methods are shown in Fig. 5.8. The calibration result of the first frame, and the reprojection errors are listed in Table 5.2.

Fig. 5.9 Reconstruction results of stone post sequence. (**a**) Three frames from the sequence, where the tracked features with relative disparities are overlaid to the last two images; (**b**) The reconstructed VRML model of the scene and the corresponding triangulated wireframe of the model shown from different viewpoints

5.6.3 Test on Medusa Head Sequence

The Medusa head video was taken on the entablature of a monumental fountain in the ancient city of Sagalassos, Turkey. The video consists of 389 frames and we select 50 frames for our test. The feature points were also detected and tracked by the system [23]. Totally 1125 features were tracked across the sequence. Some key frames, the reconstructed VRML models, and the corresponding wireframes of the two sequences are shown in Fig. 5.10. The calibration results and reprojection errors are compared in Table 5.2. The reconstructed models are realistic, though the reprojection errors are bigger than the other sequences. This is mainly due to lower image resolution and feature tracking errors encountered in long sequences.

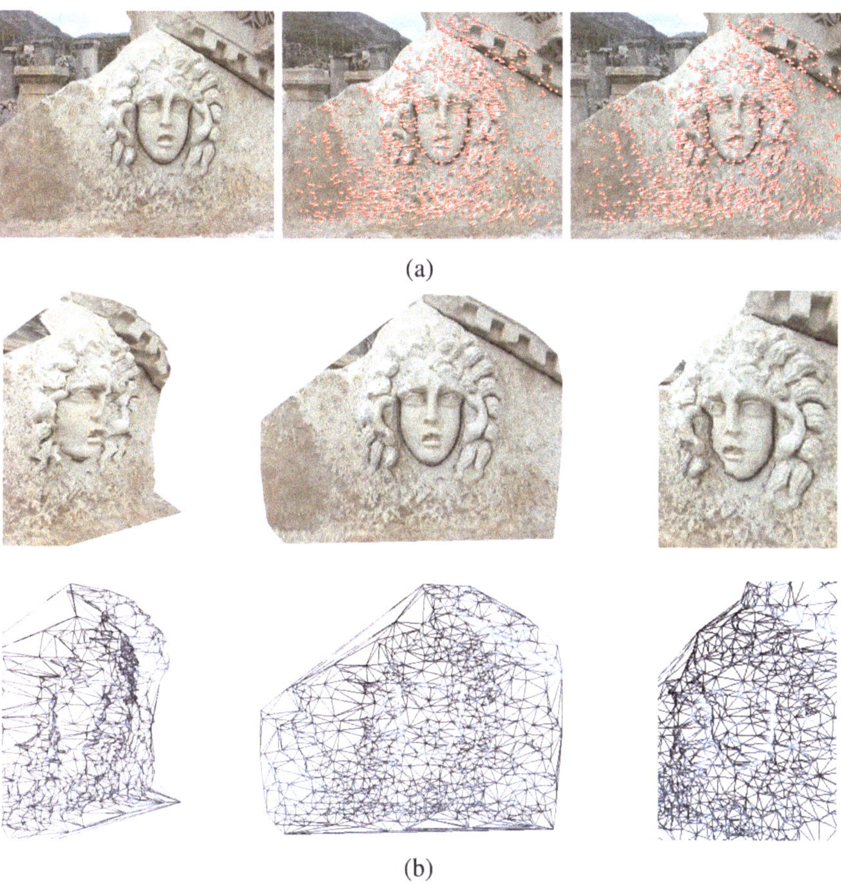

Fig. 5.10 Reconstruction results of Medusa head. (**a**) Three frames from the sequence, where the tracked features with relative disparities are overlaid to the last two images; (**b**) The reconstructed VRML model of the scene and the corresponding triangulated wireframe of the model shown from different viewpoints

5.7 Closure Remarks

5.7.1 Conclusion

In this chapter, we proposed two new algorithms to improve the overall performance of perspective factorization. One is an initialization method to recover the projective depths, which is proved to be more accurate and converges quickly than other iterative methods. The other one is a self-calibration scheme based on Kruppa constraints. It can calibrate a more general camera model, rather than just the focal length. Extensive tests and evalua-

5

tions on synthetic and real sequences demonstrated the advantages and improvements over existing methods.

5.7.2 Review Questions

1. *Projective depth initialization.* Show the principle of the proposed initialization scheme for projective depth recovery. Can you think of other ways to select two reference frames with large rotation and translation?
2. *Iterative depth recovery.* We mentioned three iteration based algorithms for projective depth recovery, i.e. Ueshiba and Tomita [22], Han and Kanade [3], and the proposed algorithm. What is the difference between them?
3. *Calibration and reconstruction.* Derive the Kruppa constraints for camera calibration. Explain the perspective factorization scheme. Show why normalization of the tracking matrix before factorization is important? How is the normalization achieved?

References

1. Christy, S., Horaud, R.: Euclidean shape and motion from multiple perspective views by affine iterations. IEEE Trans. Pattern Anal. Mach. Intell. **18**(11), 1098–1104 (1996)
2. Faugeras, O.D., Luong, Q.T., Maybank, S.J.: Camera self-calibration: Theory and experiments. In: Proc. of European Conference on Computer Vision, pp. 321–334 (1992)
3. Han, M., Kanade, T.: Creating 3D models with uncalibrated cameras. In: Proc. of IEEE Computer Society Workshop on the Application of Computer Vision (2000)
4. Hartley, R.: Kruppa's equations derived from the fundamental matrix. IEEE Trans. Pattern Anal. Mach. Intell. **19**(2), 133–135 (1997)
5. Hartley, R.I., Zisserman, A.: Multiple View Geometry in Computer Vision, 2nd edn. Cambridge University Press, Cambridge (2004). ISBN: 0521540518
6. Heyden, A., Åström, K.: Euclidean reconstruction from image sequences with varying and unknown focal length and principal point. In: Proc. of IEEE Conference on Computer Vision and Pattern Recognition, pp. 438–443 (1997)
7. Heyden, A., Åström, K.: Minimal conditions on intrinsic parameters for Euclidean reconstruction. In: Proc. of Asian Conference on Computer Vision, pp. 169–176 (1998)
8. Hu, Z., Wu, Y., Wu, F., Ma, S.D.: The number of independent Kruppa constraints from *N* images. J. Comput. Sci. Technol. **21**(2), 209–217 (2006)
9. Hung, Y., Tang, W.: Projective reconstruction from multiple views with minimization of 2D reprojection error. Int. J. Comput. Vis. **66**(3), 305–317 (2006)
10. Luong, Q., Faugeras, O.: Self-calibration of a moving camera from point correspondences and fundamental matrices. Int. J. Comput. Vis. **22**(3), 261–289 (1997)
11. Mahamud, S., Hebert, M.: Iterative projective reconstruction from multiple views. In: Proc. of IEEE Conference on Computer Vision and Pattern Recognition, vol. 2, pp. 430–437 (2000)
12. Maybank, S.: The projective geometry of ambiguous surfaces. Philos. Trans. Phys. Sci. Eng. **332**, 1–47 (1990)
13. Maybank, S., Faugeras, O.: A theory of self-calibration of a moving camera. Int. J. Comput. Vis. **8**(2), 123–151 (1992)

14. Oliensis, J., Hartley, R.: Iterative extensions of the Sturm/Triggs algorithm: Convergence and nonconvergence. IEEE Trans. Pattern Anal. Mach. Intell. **29**(12), 2217–2233 (2007)
15. Torr, P.H.S., Zisserman, A., Maybank, S.J.: Robust detection of degenerate configurations while estimating the fundamental matrix. Comput. Vis. Image Underst. **71**(3), 312–333 (1998)
16. Poelman, C., Kanade, T.: A paraperspective factorization method for shape and motion recovery. IEEE Trans. Pattern Anal. Mach. Intell. **19**(3), 206–218 (1997)
17. Pollefeys, M., Koch, R., Van Gool, L.: Self-calibration and metric reconstruction in spite of varying and unknown intrinsic camera parameters. Int. J. Comput. Vis. **32**(1), 7–25 (1999)
18. Pollefeys, M., Van Gool, L., Oosterlinck, A.: The modulus constraint: A new constraint for self-calibration. In: Proc. of International Conference on Pattern Recognition, vol. 1, pp. 349–353 (1996)
19. Quan, L.: Self-calibration of an affine camera from multiple views. Int. J. Comput. Vis. **19**(1), 93–105 (1996)
20. Sturm, P.F., Triggs, B.: A factorization based algorithm for multi-image projective structure and motion. In: Proc. of European Conference on Computer Vision, vol. 2, pp. 709–720 (1996)
21. Triggs, B.: Factorization methods for projective structure and motion. In: Proc. of the IEEE Conference on Computer Vision and Pattern Recognition, pp. 845–851 (1996)
22. Ueshiba, T., Tomita, F.: A factorization method for projective and Euclidean reconstruction from multiple perspective views via iterative depth estimation. In: Proc. of European Conference on Computer Vision, vol. 1, pp. 296–310 (1998)
23. Wang, G.: A hybrid system for feature matching based on SIFT and epipolar constraints. Tech. Rep. Department of ECE, University of Windsor (2006)
24. Wang, G., Wu, J.: Quasi-perspective projection with applications to 3D factorization from uncalibrated image sequences. In: Proc. of IEEE Conference on Computer Vision and Pattern Recognition, pp. 1–8 (2008)
25. Wang, G., Wu, J.: Perspective 3D Euclidean reconstruction with varying camera parameters. IEEE Trans. Circuits Syst. Video Technol. **19**(12), 1793–1803 (2009)
26. Wang, G., Wu, J.: Stratification approach for 3-D Euclidean reconstruction of nonrigid objects from uncalibrated image sequences. IEEE Trans. Syst. Man Cybern., Part B **38**(1), 90–101 (2008)
27. Xu, G., Sugimoto, N.: Algebraic derivation of the Kruppa equations and a new algorithm for self-calibration of cameras. J. Opt. Soc. Am. A **16**(10), 2419–2424 (1999)
28. Zaharescu, A., Horaud, R.P., Ronfard, R., Lefort, L.: Multiple camera calibration using robust perspective factorization. In: Proc. of International Symposium on 3D Data Processing, Visualization and Transmission, pp. 504–511 (2006)

6 Perspective 3D Reconstruction of Nonrigid Objects

Abstract The chapter focuses on the problem of nonrigid structure and motion factorization under perspective projection. Many previous methods are based on affine assumption that may be invalid and cause large reconstruction errors when the object is close to the camera. In this chapter, we propose two algorithms to extend these methods to full perspective projection model. The first one is a linear recursive algorithm, which updates the solution from weak-perspective to perspective projection by refining the projective depth scales. The second one is a nonlinear optimization scheme that minimizes the perspective reprojection residuals. Extensive experiments on synthetic and real image sequences are performed to validate the effectiveness of the algorithms.

The art of doing mathematics consists in finding that special case which contains all the germs of generality.

David Hilbert (1862–1943)

6.1 Introduction

The structure and motion factorization algorithm was first proposed for orthographic and weak-perspective projection models [15]. Great successes has been achieved for rigid factorization under either affine assumption [12, 14] or general perspective projection [3, 10, 18].

More generally, the factorization algorithm was extended to deal with multiple moving objects [8, 11, 25], articulated objects [17, 23], and deformable nonrigid objects [1, 2, 16]. Most of these methods assume weak-perspective camera model, which is a good approximation when the object has small depth variation and is far away from the camera. Otherwise, the recovered structure will be distorted due to perspective effect.

The extension of nonrigid factorization to perspective projection suffers from difficulties associated with perspective projection. Xiao and Kanade [22] proposed a two-step scheme to solve the problem. They first recover the projective depths iteratively using

G. Wang, Q.M.J. Wu, *Guide to Three Dimensional Structure and Motion Factorization*, Advances in Pattern Recognition,
DOI 10.1007/978-0-85729-046-5_6,

6

$3k+1$ sub-space constraints of the tracking matrix. Later the perspective structure is recovered by factorizing the scale weighted tracking matrix [21]. Del Bue *et al.* [4] proposed to segment the rigid points from the deformed ones and utilize these points to estimate camera parameters and overall motion. Perspective structure is then evaluated as a constrained nonlinear minimization with priors on the degree of deformability of each point. Lladó *et al.* [9] proposed a solution based on minimizing 2D reprojection errors, and treated the minimization problem as four weighted least-squares problems that can be solved one by one iteratively.

In this chapter, we will extend the algorithm in [3] from rigid to nonrigid case and upgrade the solution under affine assumption to that of a general perspective projection. Two algorithms are presented in the chapter. The first one is a linear recursive approximation which was proposed in [19] and the second one is a nonlinear optimization scheme [20].

The remaining part of the chapter is organized as follows: We first analyze the relationship between perspective and weak-perspective projection in Sect. 6.2. Then we present the linear recursive algorithm and the nonlinear optimization scheme in Sect. 6.3. Some experimental results on synthetic and real image sequences are given in Sects. 6.4 and 6.5 respectively.

6.2 Perspective Depth Scales and Nonrigid Factorization

In this section, we will first give an analysis of projective depth scales, then present the formulation of nonrigid factorization under weak-perspective projection.

6.2.1 Perspective Depth Scales

Suppose the images are normalized by the cameras as $\mathbf{x}_{ij} \leftarrow \mathbf{K}_i^{-1}\mathbf{x}_{ij}$. Under perspective projection, the mapping from a space point $\mathbf{X}_j$ to its normalized image $\mathbf{x}_{ij}$ in the ith frame can be expressed as

$$\rho_{ij}\mathbf{x}_{ij} = \mathbf{R}_i\bar{\mathbf{X}}_j + \mathbf{T}_i \tag{6.1}$$

where, the image point $\mathbf{x}_{ij} = [\bar{\mathbf{x}}_{ij}, 1]^T = [u_{ij}, v_{ij}, 1]^T$ is denoted in homogeneous form; $\bar{\mathbf{X}}_j$ is the inhomogeneous form of $\mathbf{X}_j$; $\mathbf{R}_i = [\mathbf{r}_{i1}^T, \mathbf{r}_{i2}^T, \mathbf{r}_{i3}^T]^T$ is the rotation matrix, and $\mathbf{T}_i = [t_{i1}, t_{i2}, t_{i3}]^T$ is the translation vector of the ith camera; ρ_{ij} is a nonzero scalar known as projective depth. Extending (6.1) we have

$$u_{ij} = \frac{\mathbf{r}_{i1}\bar{\mathbf{X}}_j + t_{i1}}{\mathbf{r}_{i3}\bar{\mathbf{X}}_j + t_{i3}} = \frac{\mathbf{r}_{i1}\bar{\mathbf{X}}_j/t_{i3} + t_{i1}/t_{i3}}{1 + \mathbf{r}_{i3}\bar{\mathbf{X}}_j/t_{i3}} = \frac{\mathbf{r}_{i1}\bar{\mathbf{X}}_j/t_{i3} + t_{i1}/t_{i3}}{1 + \varepsilon_{ij}} \tag{6.2}$$

$$v_{ij} = \frac{\mathbf{r}_{i2}\bar{\mathbf{X}}_j + t_{i2}}{\mathbf{r}_{i3}\bar{\mathbf{X}}_j + t_{i3}} = \frac{\mathbf{r}_{i2}\bar{\mathbf{X}}_j/t_{i3} + t_{i2}/t_{i3}}{1 + \mathbf{r}_{i3}\bar{\mathbf{X}}_j/t_{i3}} = \frac{\mathbf{r}_{i2}\bar{\mathbf{X}}_j/t_{i3} + t_{i2}/t_{i3}}{1 + \varepsilon_{ij}} \tag{6.3}$$

Under weak-perspective assumption, the depth variation of the object is assumed to be small compared to camera distance. This is equivalent to a zero-order approximation (i.e. $\varepsilon_{ij} = 0$) of the perspective projection. Let us denote

$$\lambda_{ij} = 1 + \varepsilon_{ij} = 1 + \mathbf{r}_{i3}\bar{\mathbf{X}}_j / t_{i3} \tag{6.4}$$

then the relationship between the images of perspective and weak-perspective projections can be expressed as:

$$\hat{\mathbf{x}}_{ij} = \lambda_{ij}\bar{\mathbf{x}}_{ij}, \quad i = 1, \ldots, m, \; j = 1, \ldots, n \tag{6.5}$$

It should be noted that we assume normalized image coordinates in the above discussion for simplicity. Nevertheless, the results (6.4) and (6.5) can be applied directly to uncalibrated images.

6.2.2 Nonrigid Affine Factorization

With Bregler's deformation model [2], the projection of a nonrigid object under weak-perspective model is given by

$$[\bar{\mathbf{x}}_{i1}, \bar{\mathbf{x}}_{i2}, \ldots, \bar{\mathbf{x}}_{in}] = \mathbf{R}_{Ai}\left(\sum_{l=1}^{k} \omega_{il}\mathbf{B}_l\right) \tag{6.6}$$

where $\mathbf{R}_{Ai}$ stands for the first two rows of the rotation matrix corresponding to the ith frame; $\mathbf{B}_l$ is the shape bases. Here the weak-perspective scaling $\alpha = f/Z_0$ is removed as it can be implicitly embedded in the deformation weights ω_{il}. Thus, weak-perspective factorization of nonrigid objects can be expressed as

$$\underbrace{\begin{bmatrix} \bar{\mathbf{x}}_{11} & \cdots & \bar{\mathbf{x}}_{1n} \\ \vdots & \ddots & \vdots \\ \bar{\mathbf{x}}_{m1} & \cdots & \bar{\mathbf{x}}_{mn} \end{bmatrix}}_{\mathbf{W}_{2m\times n}} = \underbrace{\begin{bmatrix} \omega_{11}\mathbf{R}_{A1} & \cdots & \omega_{1k}\mathbf{R}_{A1} \\ \vdots & \ddots & \vdots \\ \omega_{m1}\mathbf{R}_{Am} & \cdots & \omega_{mk}\mathbf{R}_{Am} \end{bmatrix}}_{\mathbf{M}_{2m\times 3k}} \underbrace{\begin{bmatrix} \mathbf{B}_1 \\ \vdots \\ \mathbf{B}_k \end{bmatrix}}_{\bar{\mathbf{B}}_{3k\times n}} \tag{6.7}$$

The expression is exactly the same as that under orthographic projection model. The rank of the tracking matrix is at most $3k$. We already discussed the methods of low-rank factorization, calibration, and the computation of upgrading matrix in Chap. 4. However, due to the inherent property of the affine camera model, a reversal ambiguity still remains for the recovered shapes and motions. If we denote $(\mathbf{R}_{Ai}, \bar{\mathbf{S}}_i)$ as "positive" solution of the motion and structure corresponding to ith frame, then a "negative" solution $(-\mathbf{R}_{Ai}, -\bar{\mathbf{S}}_i)$ also holds true. This ambiguity should be taken into consideration for future computations.

6

6.3 Perspective Stratification

Most previous algorithms are based on the assumption of affine camera model. This assumption becomes invalid and generate large reconstruction error when the object is close to the camera. In this section, we will introduce two algorithms to upgrade the affine solution to perspective.

6.3.1 Linear Recursive Estimation

Let us define a weighted tracking matrix by incorporating the relationship (6.5) between affine and perspective projections.

$$\dot{\mathbf{W}} = \begin{bmatrix} \hat{\mathbf{x}}_{11} & \cdots & \hat{\mathbf{x}}_{1n} \\ \vdots & \ddots & \vdots \\ \hat{\mathbf{x}}_{m1} & \cdots & \hat{\mathbf{x}}_{mn} \end{bmatrix}_{2m\times n} = \begin{bmatrix} \lambda_{11}\bar{\mathbf{x}}_{11} & \cdots & \lambda_{1n}\bar{\mathbf{x}}_{1n} \\ \vdots & \ddots & \vdots \\ \lambda_{m1}\bar{\mathbf{x}}_{m1} & \cdots & \lambda_{mn}\bar{\mathbf{x}}_{mn} \end{bmatrix}_{2m\times n} \tag{6.8}$$

If all scalars $\{\lambda_{ij} \in \mathbb{R}\}_{i=1,\ldots,nm}^{j=1,\ldots,n}$ are recovered consistently with (6.5), the shape and motion matrices obtained by factorizing the weighted tracking matrix (6.8) would correspond to a perspective projection solution. We estimate the scalars iteratively using a method similar to [3], which was originally proposed for rigid factorization. The algorithm is summarized as follows.

Algorithm: Linear recursive estimation with perspective projection. Given tracking matrix $\mathbf{W} \in \mathbb{R}^{2m\times n}$, construct the weighted tracking matrix $\dot{\mathbf{W}}$ according to (6.8) and set the initial values of $\lambda_{ij} = 1$ for $i = 1,\ldots,m$, $j = 1,\ldots,n$. Repeat the following 3 steps until λ_{ij} converges.

1. Update the weighted tracking matrix according to (6.8) and register all points in each image to its centroid;
2. Recover the Euclidean shape and motion matrices via rank-$3k$ factorization of the tracking matrix $\dot{\mathbf{W}}$;
3. Reestimate the value of λ_{ij} according to (6.4).

A theoretical proof of the convergence of this algorithm is still an open problem. However, extensive simulations show that the algorithm converges rapidly if reasonable initial values are present. Dementhon and Davis [5] and Christy and Horaud [3] also presented a qualitative analysis on the convergence of a similar recursive algorithm. In practice, one

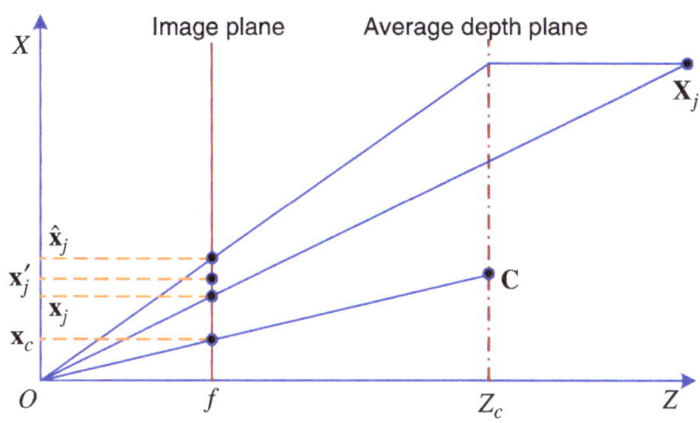

Fig. 6.1 Perspective approximation via weak-perspective projection. O is optical center of the camera, and $Z = f$ is the image plane. $\mathbf{C}$ is the centroid of the object, and $Z = Z_c$ is the average depth plane. A space point $\mathbf{X}_j$ on the object is projected to $\mathbf{x}_j$ in the image under perspective projection, and projected to $\hat{\mathbf{x}}_j$ under weak-perspective projection

convenient way to determine the convergence of the algorithm is to check the scalar variations. Let us arrange all scalars at the tth iteration in a matrix form as

$$\Lambda_t = \begin{bmatrix} \lambda_{11} & \cdots & \lambda_{1n} \\ \vdots & \ddots & \vdots \\ \lambda_{m1} & \cdots & \lambda_{mn} \end{bmatrix}_{m \times n} \tag{6.9}$$

and define the scalar variation as the Frobenius norm of the difference of the scalar matrix (6.9).

$$\delta = \left\| \Lambda_t - \Lambda_{t-1} \right\|_F \tag{6.10}$$

Experiments show that the algorithm usually converges (i.e. the scalar variation $\delta \to 0$) after 4 to 5 iterations.

Geometrical explanation of the algorithm: A geometrical explanation of the algorithm is clearly shown in Fig. 6.1. For convenience the frame subscript notation $'i'$ is omitted in the following discussion. The initial tracking matrix is composed of the image point $\mathbf{x}_j$, which is formed by perspective projection. The weighted tracking matrix of (6.8) composed of the point $\mathbf{x}'_j = \lambda_j \mathbf{x}_j$, varies in accordance with the updated scalars of step 3. Upon convergence, the image points are modified to $\hat{\mathbf{x}}_j$ such that they fit the weak-perspective projection. Thus, the algorithm recursively adjusts the image coordinates from the position of the perspective projection to that of the weak-perspective projection by adjusting the scalar λ_j. While the image of the object centroid $\mathbf{x}_c$ remains untouched during iterations.

Dealing with the reversal ambiguity: The reversal ambiguity of the factorization should be considered in the algorithm, which can be easily solved via trial and error. After recovering the shape and motion matrices in step 2, we reproject the "positive" and "negative" solutions to each frame and form two reprojected tracking matrices, say $\mathbf{W}^+$ for the "positive" and $\mathbf{W}^-$ for the "negative". Check the Frobenius norms $\|\hat{\mathbf{W}} - \mathbf{W}^+\|_F$ and $\|\hat{\mathbf{W}} - \mathbf{W}^-\|_F$. The one with smaller error would be the correct solution.

6

6.3.2 Nonlinear Optimization Algorithm

One may have noted that the solution obtained using linear recursive algorithm is just an approximation to the full perspective projection. Here we present a nonlinear optimization scheme to upgrade the reconstruction from affine to perspective projection.

Suppose the camera parameters, rotation, translation, shape bases and deformation weights under perspective projection are $\mathbf{K}_i, \mathbf{R}_i, \mathbf{T}_i, \mathbf{B}_l, \omega_{il}$, respectively. Our goal is to recover these parameters by minimizing perspective reprojection residuals defined as follows.

$$f(\mathbf{K}_i, \mathbf{R}_i, \mathbf{T}_i, \mathbf{S}_l, \omega_{il}) = \left\| \mathbf{W} - \tilde{\mathbf{W}} \right\|_F^2 = \sum_{i,j} \left\| N(\mathbf{x}_{ij}) - N(\tilde{\mathbf{x}}_{ij}) \right\|_F^2 \tag{6.11}$$

where $\tilde{\mathbf{W}}$ denotes the reprojected tracking matrix; $N(\bullet)$ denotes the normalization of a homogeneous vector so as to make its last element unity; $\tilde{\mathbf{x}}_{ij}$ is the homogeneous coordinate of the reprojected image point under perspective projection.

$$[\rho_{i1}\tilde{\mathbf{x}}_{i1}, \rho_{i2}\tilde{\mathbf{x}}_{i2}, \ldots, \rho_{in}\tilde{\mathbf{x}}_{in}] = \mathbf{R}_i \left(\sum_{l=1}^{k} \omega_{il} \mathbf{B}_l \right) + \mathbf{T}_i \tag{6.12}$$

The minimization process is also termed as bundle adjustment in computer vision society that can be solved via Newton iteration or other gradient descent methods. Here we employ the sparse Levenberg-Marquardt iteration method as given in [7]. During computation, the rotation matrix is parameterized by three parameters $\mathbf{l}_i = [l_{i1}, l_{i2}, l_{i3}]^T$ using the exponential map as

$$\mathbf{R}_i = \exp \begin{bmatrix} 0 & l_{i3} & -l_{i2} \\ -l_{i3} & 0 & l_{i1} \\ l_{i2} & -l_{i1} & 0 \end{bmatrix} \tag{6.13}$$

Please note that during each iteration, the solution is reprojected to the image via perspective projection (6.12), and the initial value may be obtained from affine factorization (6.7).

In comparison to linear recursive algorithm, the nonlinear method converges to a more accurate solution of the perspective projection, since the algorithm minimizes a geometrically meaningful cost function. Nevertheless, the nonlinear method may lead to a local minimum when initialization is poor. In practice, we can combine the two algorithms together, utilize the linear method to obtain initial values for the nonlinear scheme. We may also start with a solution from para-perspective factorization [3, 6], since this assumption is a first-order approximation of the perspective projection. The nonlinear algorithm is usually computationally intensive compared to their linear counterpart. interested readers are refereed to [3, 7] for detailed complexity comparison.

6.4 Evaluations on Synthetic Data

In this simulation, we generate a cube in a space of $20 \times 20 \times 20$ with 21 evenly distributed points on each side. The origin of the world coordinate system is set at the center of the cube. There are three sets of points (33×3 points) on the adjacent three surfaces of the cube that move outward along the coordinate axes at constant speed toward outside as shown in Fig. 6.2. We generate 30 cubes in space with randomly selected poses, and project each cube to an image by perspective projection. Altogether there are 351 image points in each frame, amongst which 252 points belong to the cube and the remaining 99 points belong

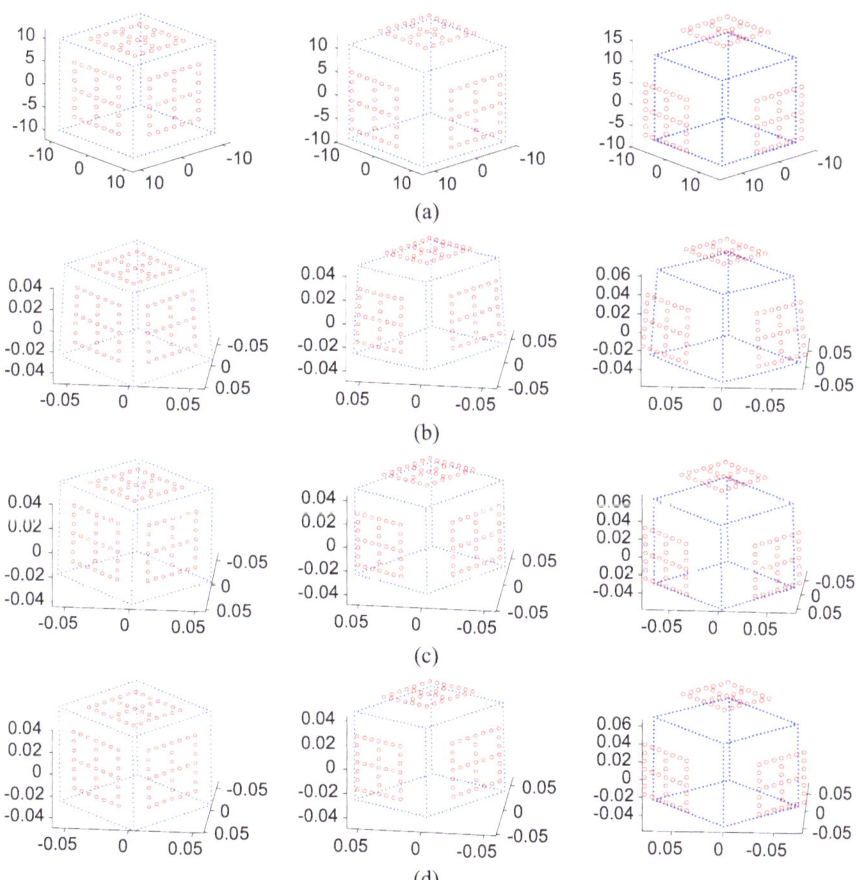

Fig. 6.2 Synthetic data and reconstruction results by different algorithms. (**a**) Three generated 3D shapes of the synthetic cubes (*in dot*) together with the three sets of moving points (*in circle*); (**b**) The corresponding recovered structures by weak-perspective factorization; (**c**) The upgraded perspective shapes by the linear recursive algorithm; (**d**) The upgraded structures by the nonlinear optimization algorithm

to the three moving parts. During shooting, the distance of the camera to the object is set at about 14 times of the object size such that the imaging condition is quite close to weak-perspective assumption.

6.4.1 Reconstruction Results

We first recover shape and motion matrices corresponding to each frame using the factorization algorithm under affine assumption. Then upgrade the solution from affine to perspective projection according to the two proposed algorithms. Figure 6.2 shows the recovered 3D structures of three frames by different algorithms. It is evident from the results that both the linear recursive and the nonlinear algorithm achieve very good results, and are quite similar to the ground truths visually. The perspective distortion of affine factorization is obvious in Fig. 6.2, the reconstructed structures are not exactly cubes and the reconstructed lines are curved due to perspective effect, even though the camera setup in the simulation is very close to weak-perspective assumption.

One may have noted that each reconstructed shape in Fig. 6.2 is defined up to a 3D similarity transformation with the ground truth. For ease of evaluation, we first compute the transformation matrix by virtue of point correspondences between the recovered structure and its ground truth, then transform each reconstructed cube to the coordinate frame of its associated ground truth. We define the reconstruction errors as point-to-point distances between the reconstructed structure and its ground truth. Table 6.1 shows the mean and standard deviation (STD) of the reconstruction errors of the two frames, where 'Affine' stands for the weak-perspective solution, 'Linear' refers to the linear recursive algorithm and 'Nonlinear' denotes the nonlinear optimization algorithm respectively. We infer from Table 6.1 that the structures recovered by the two proposed algorithms are more accurate than those by affine assumption.

6.4.2 Convergence and Performance Comparisons

We evaluated the convergence property of the two proposed algorithms. For all iterations, we record the scalar variation (6.10) and the relative reprojection error defined as

Table 6.1 Evaluation of reconstruction errors. We register the recovered structures with respect to the ground truths, then compare the mean and standard deviation of the distances between the reconstruction and the associated ground truth

Frame number	Frame 1			Frame 10		
Algorithm	Affine	Linear	Nonlinear	Affine	Linear	Nonlinear
Mean	0.3580	0.0135	0.0128	0.3624	0.0198	0.0186
STD	0.0961	0.0078	0.0067	0.1026	0.0091	0.0082

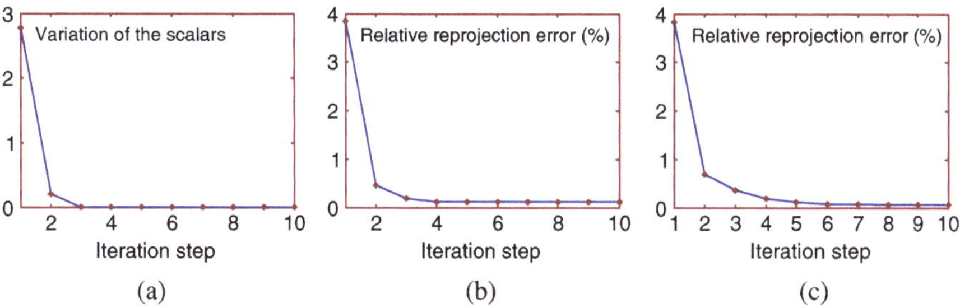

Fig. 6.3 Performance of the two upgrading algorithms. (**a**) The scalar variations of the linear recursive algorithm with respect to iteration steps; (**b**) The relative reprojection error of the linear recursive algorithm at each iteration; (**c**) The relative reprojection error of the nonlinear optimization algorithm at each iteration

follows.

$$\mathscr{E}_t = \frac{\|\mathbf{W} - \mathbf{W}_t\|_F}{\|\mathbf{W}\|_F} \times 100\ (\%) \tag{6.14}$$

where, $\mathbf{W}$ is the initial tracking matrix, $\mathbf{W}_t$ is the reprojected tracking matrix at the tth iteration. The results are shown in Fig. 6.3.

It is evident from Fig. 6.3 that the convergence speed of the two algorithms is very fast. The reprojection error at first iteration corresponds to the weak-perspective solution, and the error is significantly reduced by the proposed upgrading algorithms. There is no added noise in the test, however, a small residual error still exists after convergence since the images are produced by full perspective projection rather than the weak-perspective projection. It is obvious that the residual error will increase with an increase of in noise level. For 10 iterations, the nonlinear algorithm requires only 158.3 seconds, while the linear recursive method only takes 11.2 seconds on a Dell PC with Intel 2.6 GHz CPU programmed with Matlab 6.5.

We studied the performance of different algorithms with respect to the relative distance which is defined as the ratio of the distance from object to the camera over the object depth. We utilize the same synthetic data and vary the relative distance from 7 to 14 in steps of 1 and produce different image sets. For each set of images, we recover the affine structure and motion of each frame, and upgrade the solution to perspective projection by the two proposed algorithms. As evaluation, we first calculate the relative reprojection error according to (6.14), then register the reconstructed structure with its ground truth and check the reconstruction error. The mean and standard deviation of the error corresponding to the 5th frame are shown in Table 6.2.

Table 6.2 shows that the reconstruction errors increase as the camera moves closer to the object. The two proposed upgrading algorithms are initialized by weak-perspective solution, thus they may not converge to the correct solution when initial value error increase beyond a certain extent. As shown in Table 6.2, when the relative distance is set at 7,

Table 6.2 Performance comparison of different algorithms with respect to the relative distances from the object to the cameras. 'Error' stands for the relative reprojection error, 'Iteration' stands for the iteration times of convergence

Relative distance		7	8	9	10	11	12	13	14
	Affine	8.264	6.043	5.436	4.957	4.582	4.282	4.035	3.824
Error (%)	Linear	1.267	0.319	0.256	0.219	0.199	0.176	0.169	0.154
	Nonlinear	1.258	0.314	0.251	0.215	0.192	0.170	0.162	0.149
	Affine	0.713	0.623	0.551	0.490	0.449	0.416	0.384	0.362
Mean	Linear	1.069	0.042	0.036	0.034	0.031	0.027	0.018	0.015
	Nonlinear	1.066	0.040	0.035	0.023	0.030	0.025	0.017	0.015
	Affine	0.169	0.146	0.133	0.122	0.112	0.107	0.102	0.099
STD	Linear	0.167	0.090	0.064	0.037	0.027	0.013	0.010	0.008
	Nonlinear	0.166	0.081	0.063	0.035	0.025	0.011	0.009	0.007
Iteration	Linear	6	6	5	5	5	5	5	5
	Nonlinear	8	7	7	6	6	6	6	6

the two algorithms converge in 6 and 8 iterations respectively. However, they converge to false solutions due to bad initialization. Generally speaking, Reasonable solutions can be guaranteed when the relative distance is greater than 8. We also find that the nonlinear method is more sensitive to initial values than the linear recursive algorithm.

6.5 Experiments with Real Sequences

We tested the proposed methods on several real image sequences and report results on two sequences in this section.

6.5.1 Test on Franck Sequence

The sequence is downloaded from the European working group on face and gesture recognition (www-prima.inrialpes.fr/fgnet/) with a resolution of 720×576. We select 60 frames with various facial expressions for the experiment. The tracking data, which contain 68 automatically tracked feature points using the active appearance model (AAM) method, are also downloaded from the group. In this test, we adopt a simplified camera model with square pixel and the principal point at the image center, the intrinsic parameters of the camera are estimated by the method of [13] with rigid approximation.

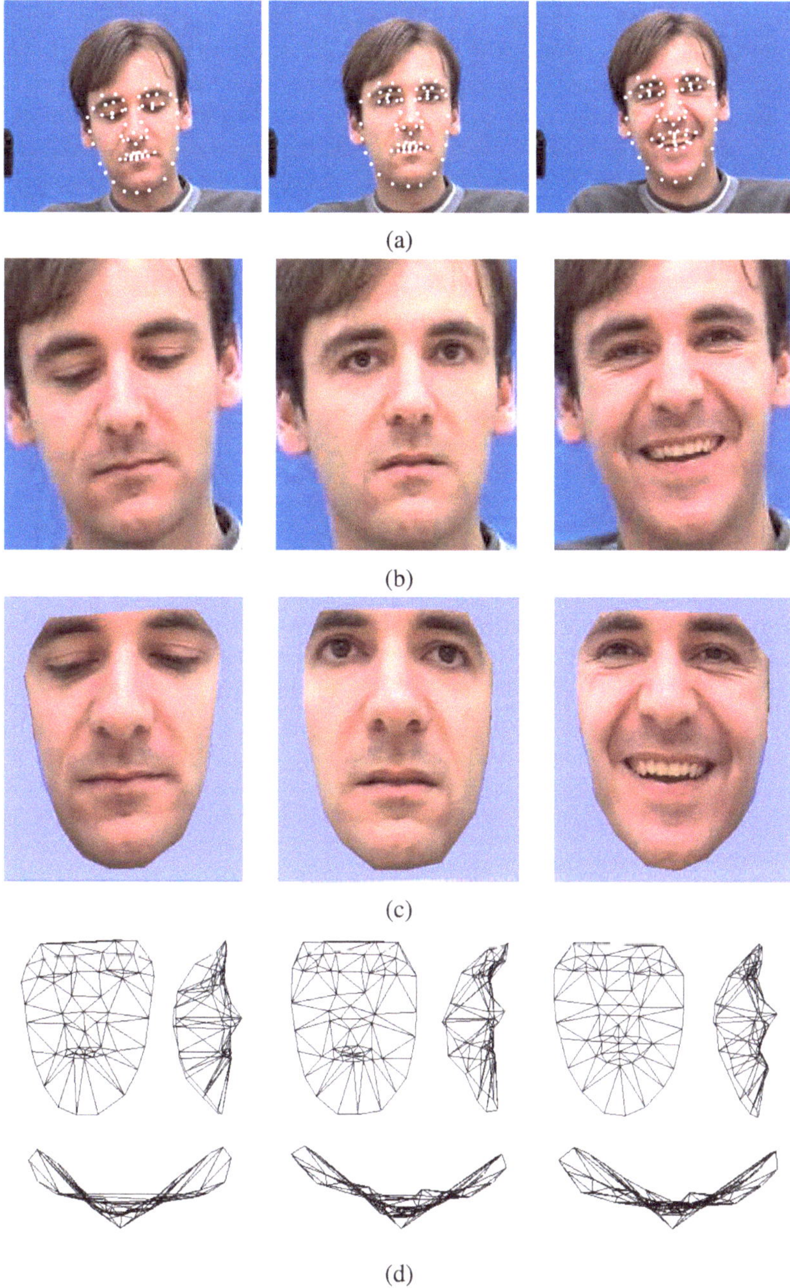

Fig. 6.4 Reconstruction results of five frames with different facial expressions from Franck sequence. (**a**) Five key frames overlaid with 68 tracked feature points; (**b**) The enlarged faces of the five frames; (**c**) The front views of the five reconstructed VRML models with texture mapping; (**d**) The front, side and top views of the corresponding triangulated wireframe models

Figure 6.4 shows the reconstructed VRML models with texture mapping and the triangulated wireframes corresponding to five frames, which are obtained by the linear recursive algorithm with 10 iterations. The reconstruction results by other methods are omitted here, since there is no much visual difference. The models associated with different frames and different facial expressions are correctly recovered. The results could be used for visualization and recognition. However, the positions of some reconstructed features are not accurate due to errors in tracked features and camera parameters. For a comparison, the relative reprojection errors of the weak perspective factorization, the linear recursive and the nonlinear optimization algorithm are 6.42%, 0.27%, 0.21%, respectively. The improvements of the proposed algorithms are pretty obvious.

6.5.2 Test on Scarf Sequence

The sequence is taken by a Canon PowerShot G3 with a resolution of 1024×768. The camera is pre-calibrated by Zhang's method [26]. There are 15 frames in the sequence and the scarf is pressed during shooting so as to deform it. We utilize the method proposed by Yao and Cham [24] to seek correspondences between these frames and delete outliers interactively. Totally 2986 features are tracked across the sequence as shown in Fig. 6.5. We recover the affine shape and motion matrices of each frame by weak-perspective factorization, then upgrade the solution to a full perspective projection by the linear recursive algorithm. Figure 6.5 shows the reconstructed VRML models and wireframes corresponding to three frames of the sequence. The recovered 3D structures of the scarf with deformations are visually plausible and realistic.

As a comparison, the relative reprojection errors of the three algorithms are compared in Table 6.3. The background consists of two orthogonal sheets with square grids. We take this as a ground truth of evaluation, and calculate the reconstructed angle between the two sheets of each frame. The mean of the errors by the three algorithms are listed in Table 6.3. There are altogether 24 reconstructed square grids in the sequence, we check the length ratio of the two diagonals of each square and the angle formed by the two diagonals. The mean of the length ratios and the angles formed are shown in Table 6.3, noticeable improvements over previous affine solution are evident.

Table 6.3 Evaluation results of the recovered structures associated with the scarf sequence

	Reprojection error	Mean error of reconstructed angle	Mean error of length ratios	Mean error of diagonal angles
Affine	4.946%	3.455°	0.126	0.624°
Linear	0.409%	1.736°	0.085	0.338°
Nonlinear	0.342%	1.384°	0.079	0.325°

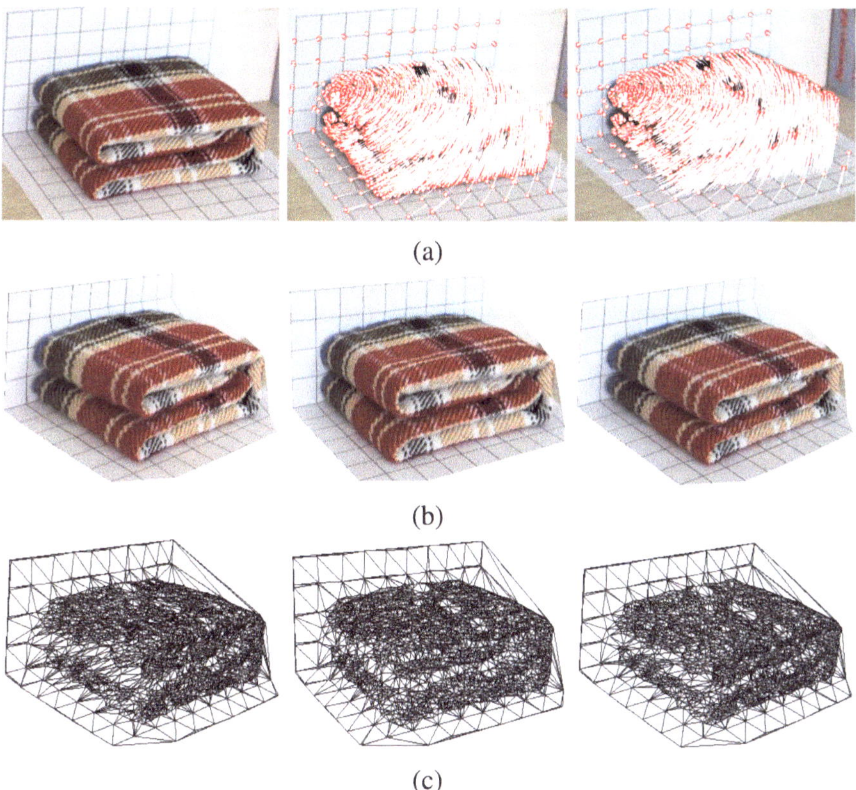

Fig. 6.5 Reconstruction results of the scarf sequence. (a) Three frames of the sequence with 2986 tracked features overlaid to two frames with relative disparities shown in *white lines*; (b) The reconstructed VRML models of the scarf shown in different viewpoints with texture mapping; (c) The corresponding wireframes of the VRML models

6.6 Closure Remarks

6.6.1 Conclusion

In this chapter, we introduced two algorithms to update previous nonrigid factorization result from affine to full perspective camera model. The algorithms show good improvements over previous solutions. Both algorithms are based on the initialization of affine factorization. The key point of the linear recursive algorithm is to recover the projective depths. A related study can be found in [22], and more similar works were discussed in Chap. 5. The nonlinear algorithm can be used as a second step of the linear recursive estimation to further refine the solution.

6

6.6.2 Review Questions

1. *Projective depth scale.* Derive the relationship (6.5) that exists between perspective and weak-perspective projection for uncalibrated images.
2. *Linear recursive estimation.* Show how projective depth scales in each iteration are updated. Can you think of other ways to update the depth scales? Describe the geometrical meaning of the algorithm.
3. *Nonlinear optimization algorithm.* During each iteration step of the optimization process. Explain how the solution is back-projected to images by perspective projection?

References

1. Brand, M.: A direct method for 3D factorization of nonrigid motion observed in 2D. In: Proc. of IEEE Conference on Computer Vision and Pattern Recognition, vol. 2, pp. 122–128 (2005)
2. Bregler, C., Hertzmann, A., Biermann, H.: Recovering non-rigid 3D shape from image streams. In: Proc. of IEEE Conference on Computer Vision and Pattern Recognition, vol. 2, pp. 690–696 (2000)
3. Christy, S., Horaud, R.: Euclidean shape and motion from multiple perspective views by affine iterations. IEEE Trans. Pattern Anal. Mach. Intell. **18**(11), 1098–1104 (1996)
4. Del Bue, A., Lladó, X., de Agapito, L.: Non-rigid metric shape and motion recovery from uncalibrated images using priors. In: Proc. of IEEE Conference on Computer Vision and Pattern Recognition, vol. 1, pp. 1191–1198 (2006)
5. Dementhon, D.F., Davis, L.S.: Model-based object pose in 25 lines of code. Int. J. Comput. Vis. **15**(1–2), 123–141 (1995)
6. Fujiki, J., Kurata, T.: Recursive factorization method for the paraperspective model based on the perspective projection. In: Proc. of International Conference on Pattern Recognition, vol. 1, pp. 406–410 (2000)
7. Hartley, R.I., Zisserman, A.: Multiple View Geometry in Computer Vision, 2nd edn. Cambridge University Press, Cambridge (2004). ISBN: 0521540518
8. Li, T., Kallem, V., Singaraju, D., Vidal, R.: Projective factorization of multiple rigid-body motions. In: Proc. of IEEE Conference on Computer Vision and Pattern Recognition (2007)
9. Lladó, X., Del Bue, A., de Agapito, L.: Non-rigid 3D factorization for projective reconstruction. In: Proc. of British Machine Vision Conference (2005)
10. Oliensis, J., Hartley, R.: Iterative extensions of the Sturm/Triggs algorithm: Convergence and nonconvergence. IEEE Trans. Pattern Anal. Mach. Intell. **29**(12), 2217–2233 (2007)
11. Ozden, K.E., Schindler, K., Van Gool, L.: Multibody structure-from-motion in practice. IEEE Trans. Pattern Anal. Mach. Intell. **32**(6), 1134–1141 (2010)
12. Poelman, C., Kanade, T.: A paraperspective factorization method for shape and motion recovery. IEEE Trans. Pattern Anal. Mach. Intell. **19**(3), 206–218 (1997)
13. Pollefeys, M., Koch, R., Van Gool, L.: Self-calibration and metric reconstruction in spite of varying and unknown intrinsic camera parameters. Int. J. Comput. Vis. **32**(1), 7–25 (1999)
14. Quan, L.: Self-calibration of an affine camera from multiple views. Int. J. Comput. Vis. **19**(1), 93–105 (1996)
15. Tomasi, C., Kanade, T.: Shape and motion from image streams under orthography: a factorization method. Int. J. Comput. Vis. **9**(2), 137–154 (1992)
16. Torresani, L., Hertzmann, A., Bregler, C.: Nonrigid structure-from-motion: Estimating shape and motion with hierarchical priors. IEEE Trans. Pattern Anal. Mach. Intell. **30**(5), 878–892 (2008)

17. Tresadern, P., Reid, I.: Articulated structure from motion by factorization. In: Proc. of the IEEE Conference on Computer Vision and Pattern Recognition, vol. 2, pp. 1110–1115 (2005)
18. Triggs, B.: Factorization methods for projective structure and motion. In: Proc. of the IEEE Conference on Computer Vision and Pattern Recognition, pp. 845–851 (1996)
19. Wang, G., Tian, Y., Sun, G.: Modelling nonrigid object from video sequence under perspective projection. In: Lecture Notes in Computer Science, vol. 3784, pp. 64–71 (2005)
20. Wang, G., Tsui, H.T., Hu, Z.: Structure and motion of nonrigid object under perspective projection. Pattern Recogn. Lett. **28**(4), 507–515 (2007)
21. Xiao, J., Chai, J.X., Kanade, T.: A closed-form solution to non-rigid shape and motion recovery. In: Proc. of European Conference on Computer Vision, vol. 4, pp. 573–587 (2004)
22. Xiao, J., Kanade, T.: Uncalibrated perspective reconstruction of deformable structures. In: Proc. of the International Conference on Computer Vision, vol. 2, pp. 1075–1082 (2005)
23. Yan, J., Pollefeys, M.: A factorization-based approach for articulated nonrigid shape, motion and kinematic chain recovery from video. IEEE Trans. Pattern Anal. Mach. Intell. **30**(5), 865–877 (2008)
24. Yao, J., Cham, W.: Feature matching and scene reconstruction from multiple widely separated views. Tech. Rep., Chinese University of Hong Kong (2005)
25. Zelnik-Manor, L., Machline, M., Irani, M.: Multi-body factorization with uncertainty: Revisiting motion consistency. Int. J. Comput. Vis. **68**(1), 27–41 (2006)
26. Zhang, Z.: A flexible new technique for camera calibration. IEEE Trans. Pattern Anal. Mach. Intell. **22**(11), 1330–1334 (2000)

7 Rotation Constrained Power Factorization

Abstract The chapter addresses an alternative method for structure and motion factorization of nonrigid objects. We present a rotation constrained power factorization (RCPF) algorithm that integrates orthonormality and replicated block structure of the motion matrix directly into iterations. The algorithm is easy to implement and can work with incomplete tracking matrix. Based on the shape bases recovered by the batch-type factorization, we introduce a sequential factorization technique to compute the shape and motion of new frames efficiently. Extensive experiments show the effectiveness of the proposed algorithm.

We must confine ourselves to those forms that we know how to handle, or for which any tables which may be necessary have been constructed.

Sir Ronald Aylmer Fisher (1890–1962)

7.1 Introduction

Most structure and motion factorization algorithms for both rigid objects [8, 18] and nonrigid objects [2, 4] are based on SVD decomposition. Original SVD based algorithms can not work with a tracking matrix with some missing entries. While in practice, it is hard to have all features tracked across the entire sequence due to self-occlusion, varying illumination, and other reasons. It is common to require operations on matrix with missing data.

Brand [3] proposed an incremental SVD to efficiently deal with and update the missing values. Martinec and Pajdla [16] solved the missing data problem under perspective factorization framework [20]. Gruber and Weiss [11, 12] proposed an expectation maximization (EM) algorithm to perform factorization with missing data and uncertainties. Hung and Tang [14] integrated initial search and bundle adjustment into a single algorithm that consists of a sequence of weighted least-square problem. Jacobs [15] treated each column with missing entries as an affine subspace and solved the problem by obtaining the intersection of all quadruple affine subspaces. Chen and Suter [7] developed a criterion to recover the

G. Wang, Q.M.J. Wu, *Guide to Three Dimensional Structure and Motion Factorization*, Advances in Pattern Recognition,
DOI 10.1007/978-0-85729-046-5_7,

most reliable submatrix in presence of missing values and applied it to the problem of missing data in a large low-rank matrix. Torresani *et al.* [22] proposed a probabilistic principal components analysis algorithm to estimate the 3D shape and motion with missing data. There are some other nonlinear based studies to deal with incomplete tracking matrix, such as damped Newton method [5] and Levenberg-Marquardt based method [6].

The essence of the factorization algorithm is to find a low-rank approximation of the tracking matrix. Different with previous SVD-based methods, Hartley and Schaffalitzky [13] proposed an alternate algorithm, named power factorization (PF), to find a low-rank approximation of the tracking matrix. The method is derived from the power method in matrix computation [10] and the sequential factorization method proposed by Morita and Kanade [17]. Vidal and Hartley [24] proposed to combine the PF algorithm for motion segmentation. The PF algorithm is easy to implement and can handle missing data in a tracking matrix.

All the above methods were initially designed for rigid objects and static scenes. Wang *et al.* extended the PF algorithm to nonrigid factorization in both metric space [26] and affine space [27]. In this chapter, we will introduce the rotation constrained power factorization (RCPF) algorithm for nonrigid objects [26]. The remaining part of the chapter is organized as follows: The general power factorization algorithm is briefly reviewed in Sect. 7.2. The RCPF algorithm is elaborated in Sect. 7.3. Some experimental evaluations on synthetic and real images are presented in Sects. 7.4 and 7.5 respectively.

7.2 Power Factorization for Rigid Objects

Given the tracked features $\{\bar{\mathbf{x}}_{ij} \in \mathbb{R}^2\}$, $i = 1, \ldots, m$, $j = 1, \ldots, n$, across a sequence of m video frames of a rigid object. Under affine projection model, the factorization expression can be written as

$$\begin{bmatrix} \bar{\mathbf{x}}_{11} & \cdots & \bar{\mathbf{x}}_{1n} \\ \vdots & \ddots & \vdots \\ \bar{\mathbf{x}}_{m1} & \cdots & \bar{\mathbf{x}}_{mn} \end{bmatrix} = \begin{bmatrix} \mathbf{A}_1 \\ \vdots \\ \mathbf{A}_m \end{bmatrix} [\bar{\mathbf{X}}_1, \ldots, \bar{\mathbf{X}}_n] \tag{7.1}$$

This set of equations may be written as $\mathbf{W} = \mathbf{A}\mathbf{B}^T$. The problem of affine factorization is exactly the problem of factoring tracking matrix $\mathbf{W}$ into two rank-3 factors. In presence of noise, the rank of the tracking matrix is greater than 3. Thus we will find the closest rank-3 approximation $\tilde{\mathbf{W}} = \mathbf{A}\mathbf{B}^T$, so as to minimize the Frobenius norm of the difference.

$$J_1 = \min_{\mathbf{A},\mathbf{B}} \|\mathbf{W} - \tilde{\mathbf{W}}\|_F = \min_{\mathbf{A},\mathbf{B}} \|\mathbf{W} - \mathbf{A}\mathbf{B}^T\|_F \tag{7.2}$$

where $\mathbf{A}$ and $\mathbf{B}$ have just three columns. Assuming isotropic IID Gaussian image noise, $\tilde{\mathbf{W}}$ will be the maximum likelihood estimation of the tracking matrix. Most researchers carry out the minimization by SVD decomposition of $\mathbf{W}$ and truncating all but its first three singular values to zero [8, 18, 21].

More generally, suppose $r < \min(2m, n)$, we wish to find two rank-r matrices $\mathbf{A} \in \mathbb{R}^{2m\times r}$ and $\mathbf{B} \in \mathbb{R}^{n\times r}$ such that $\|\mathbf{W} - \mathbf{A}\mathbf{B}^T\|_F$ is minimized. Based on the power method [10], Hartley and Schaffalitzky [13] proposed an alternative power factorization (PF) algorithm to find a low-rank approximation of a tracking matrix. Starting with an initial random $2m \times r$ matrix $\mathbf{A}_0$, They iterate the following two steps until the product $\mathbf{A}_t\mathbf{B}_t^T$ converges.

$$\begin{aligned} \mathbf{B}_t &= \mathbf{W}^T\mathbf{A}_{t-1}\mathbf{N}_t \\ \mathbf{A}_t &= \mathbf{W}\mathbf{B}_t\left(\mathbf{B}_t^T\mathbf{B}_t\right)^{-1} \end{aligned} \tag{7.3}$$

where $\mathbf{N}_t$ is a nonsingular $r \times r$ normalization matrix that offers numerical stability to the algorithm. A different form of normalization matrix can be made in which the two steps are symmetric.

$$\begin{aligned} \mathbf{B}_t &= \left(\mathbf{W}^T\mathbf{A}_{t-1}\right)\left(\mathbf{A}_{t-1}^T\mathbf{A}_{t-1}\right)^{-1} \\ \mathbf{A}_t &= \left(\mathbf{W}\mathbf{B}_t\right)\left(\mathbf{B}_t^T\mathbf{B}_t\right)^{-1} \end{aligned} \tag{7.4}$$

Compared with the SVD-based method, the PF algorithm is very simple and computationally cheap, since it only involves the multiplication and inverse of matrices. As seen from (7.4), each step is exactly a least-squares solution to a set of equations of the form $\mathbf{W} = \mathbf{A}\mathbf{B}^T$.

Computationally, the updates of $\mathbf{B}$ given $\mathbf{A}$, or vice versa, can be done by solving a least squares problem for each row of $\mathbf{B}$ or $\mathbf{A}$ independently. This allows us to deal with missing entries in the tracking matrix. Missing entries in $\mathbf{W}$ correspond to omitted equations.

7.3 Power Factorization for Nonrigid Objects

Under weak perspective assumption, a given tracking matrix $\mathbf{W}$ of a nonrigid object can be factorized into the following motion and shape matrices [4].

$$\mathbf{W}_{2m\times n} = \mathbf{M}_{2m\times 3k}\bar{\mathbf{B}}_{3k\times n} \tag{7.5}$$

where

$$\mathbf{M} = \begin{bmatrix} \omega_{11}\mathbf{R}_{A1} & \cdots & \omega_{1k}\mathbf{R}_{A1} \\ \vdots & \ddots & \vdots \\ \omega_{m1}\mathbf{R}_{Am} & \cdots & \omega_{mk}\mathbf{R}_{Am} \end{bmatrix}, \qquad \bar{\mathbf{B}} = \begin{bmatrix} \mathbf{B}_1 \\ \vdots \\ \mathbf{B}_k \end{bmatrix} \tag{7.6}$$

In case of nonrigid factorization, our objective is to recover the motion matrix $\mathbf{M} \in \mathbb{R}^{2m\times 3k}$ and the shape matrix $\bar{\mathbf{B}} \in \mathbb{R}^{3k\times n}$. It appears that the general power factorization algorithm (7.4) may be applied directly to recover the shape and motion matrices. However, the solution does not observe the replicated block structure of the motion matrix in (7.6), thus it is defined up to a correction matrix $\mathbf{H}$ that is difficult to compute as

7

discussed in Chap. 4. We will introduce a rotation constrained power factorization (RCPF) algorithm to solve this problem.

7.3.1 Rotation Constrained Power Factorization

Let us decompose the motion matrix (7.6) into the following two parts.

$$\mathbf{M} = \Omega \otimes \Phi = \begin{bmatrix} \omega_{11}\mathbf{E} & \cdots & \omega_{1k}\mathbf{E} \\ \vdots & \ddots & \vdots \\ \omega_{m1}\mathbf{E} & \cdots & \omega_{mk}\mathbf{E} \end{bmatrix} \otimes \begin{bmatrix} \mathbf{R}_{A1} & \cdots & \mathbf{R}_{A1} \\ \vdots & \ddots & \vdots \\ \mathbf{R}_{Am} & \cdots & \mathbf{R}_{Am} \end{bmatrix} \tag{7.7}$$

where Ω denotes the weighting matrix, Φ denotes the rotation matrix, $\mathbf{E}$ is a 2×3 matrix with unit entries, '$\otimes$' denotes the Hadamard product of element-by-element multiplication. Let us define a cost function

$$J_2 = \min_{\Omega, \Phi, \bar{\mathbf{B}}} \left\| \mathbf{W} - (\Omega \otimes \Phi)\bar{\mathbf{B}} \right\|_F^2 \tag{7.8}$$

The algorithm can be summarized as follows.

Algorithm: Rotation constrained power factorization. Given tracking matrix $\mathbf{W} \in \mathbb{R}^{2m \times n}$, initial rotation matrix $\Phi \in \mathbb{R}^{2m \times 3k}$, and initial shape matrix $\bar{\mathbf{B}} \in \mathbb{R}^{3k \times n}$. Repeat the following three steps until the convergence of $(\Omega \otimes \Phi)\bar{\mathbf{B}}$.

1. Update Ω by minimizing (7.8) with given Φ and $\bar{\mathbf{B}}$;
2. Update $\bar{\mathbf{B}}$ by minimizing (7.8) with given Ω and Φ;
3. Update Φ by minimizing (7.8) with given Ω and $\bar{\mathbf{B}}$, subject to the constraint that Φ is a block replicated rotation matrix.

The main difference between RCPF algorithm and the general power factorization algorithm is that we divide the computation of $\mathbf{M}$ into Ω and Φ two steps. This makes it possible to explicitly combine the orthonormal property of the rotation matrix and the replicated block structure of the motion matrix into the minimization scheme [25]. The computational details involved in each step are described below.

Step 1. Suppose $\mathbf{W}_i \in \mathbb{R}^{2 \times n}$ is the ith two-row of $\mathbf{W}$, which is composed of the tracked features in the ith frame, then we have:

$$\mathbf{W}_i = \mathbf{M}_i \bar{\mathbf{B}} = [\omega_{i1}\mathbf{R}_{Ai}, \ldots, \omega_{ik}\mathbf{R}_{Ai}] \begin{bmatrix} \mathbf{B}_1 \\ \vdots \\ \mathbf{B}_k \end{bmatrix} \tag{7.9}$$

where $\mathbf{R}_{Ai}$ represents the first two-rows of the rotation matrix corresponding to the ith frame, $\mathbf{B}_l$ is the shape basis. The motion matrix can be easily solved from (7.9) in least-squares as

$$\mathbf{M}_i = \mathbf{W}_i \bar{\mathbf{B}}^T \left(\bar{\mathbf{B}}\bar{\mathbf{B}}^T\right)^{-1} \tag{7.10}$$

Thus, the deformation weight ω_{il} is factorized using (7.10). The solution may be further optimized by minimizing

$$f(\omega_{il}) = \min \left\| \mathbf{W}_i - \sum_{l=1}^{k} \omega_{il} \mathbf{R}_{Ai} \mathbf{B}_l \right\|_F^2 \tag{7.11}$$

Step 2. After obtaining the deformation weights in step 1, the weighting matrix Ω can be updated accordingly. Then the shape matrix is easily updated from $\mathbf{W} = (\Omega \otimes \Phi)\bar{\mathbf{B}}$ by least-squares.

$$\bar{\mathbf{B}} = \left((\Omega \otimes \Phi)^T (\Omega \otimes \Phi)\right)^{-1} \left((\Omega \otimes \Phi)^T \mathbf{W}\right) \tag{7.12}$$

Step 3. From the updated shape matrix in step 2, we update the structure associated with each frame using

$$\bar{\mathbf{S}}_i = \sum_{l=1}^{k} \omega_{il} \mathbf{B}_l \tag{7.13}$$

Let us rewrite (7.9) as $\mathbf{W}_i = \mathbf{R}_{Ai}\bar{\mathbf{S}}_i$. Then the rotation matrix can be simply computed by least-squares.

$$\mathbf{R}_{Ai} = \mathbf{W}_i \left(\bar{\mathbf{S}}_i\right)^T \left(\bar{\mathbf{S}}_i \bar{\mathbf{S}}_i^T\right)^{-1} \tag{7.14}$$

Usually, the recovered matrix in (7.14) is not an orthonormal rotation matrix due to image noise. Suppose the SVD decomposition of $\mathbf{R}_{Ai}$ is $\mathbf{U}\Sigma\mathbf{V}^T$, where $\mathbf{U}$ and $\mathbf{V}$ are orthogonal matrices in the dimension of 2×2 and 3×3 respectively, Σ is a 2×3 diagonal matrix composed by the two singular values of $\mathbf{R}_{Ai}$. Thus the best approximation to the rotation matrix can be obtained from

$$\mathbf{R}_{Ai} = \mathbf{U} \begin{bmatrix} 1 & 0 & 0 \\ 0 & 1 & 0 \end{bmatrix} \mathbf{V}^T \tag{7.15}$$

since a rotation matrix should have unit singular values. Then the block replicated rotation matrix Φ can be resembled from (7.15).

As one may have noticed, each minimization step in the algorithm is equivalent to solving a set of equations by least-squares. This allows us to handle the tracking matrix with some missing entries as that in general power factorization [13]. In case of missing data, the cost function (7.8) is modified to

$$J_3 = \sum_{(i,j)\in\Xi} \left| W_{ij} - \left((\Omega \otimes \Phi)\bar{\mathbf{B}}\right)_{ij} \right|^2 \tag{7.16}$$

where W_{ij} denotes the (i, j)th element of the tracking matrix, Ξ stands for the set of available entries in the tracking matrix. Thus we update Ω, $\bar{\mathbf{B}}$, and Φ using the available features in $\mathbf{W}$ according to (7.16). This is a very important attribute of the algorithm, since it is hard to track all features across the whole sequence due to self-occlusion in real applications.

7.3.2 Initialization and Convergence Determination

There is currently no theoretical proof of the convergence of the power factorization algorithm [13]. Nevertheless, through extensive simulation tests, we find that the algorithm converges quickly to a correct solution when the rank of $\mathbf{W}$ is close to $3k$ and reasonable initial values are present. Specifically, suppose σ_i is the ith largest singular value of $\mathbf{W}$, then the convergence is proportional to $(\sigma_{3k+1}/\sigma_{3k})^{2t}$. Compared to general power factorization, the convergence speed of the RCPF algorithm is somewhat slower due to enforcement of orthonormal constraint on the rotation matrix in step 3.

There are several possible ways to determine convergence of the algorithm. The most feasible way is to check the variation of the reprojected tracking matrix which is defined as the Frobenius norm of the difference with the last iteration.

$$\delta = \left\| \mathbf{W}_t - \mathbf{W}_{t-1} \right\|_F^2 \tag{7.17}$$

where $\mathbf{W}_t = (\Omega_t \otimes \Phi_t)\bar{\mathbf{B}}_t$ is the reprojected tracking matrix at the tth iteration. As demonstrated in the experiments, the algorithm usually converges after 5 iterations.

For nonrigid factorization, the expected structure of the object is expressed as the weighted combination of a set of shape bases. The expression is not unique if no constraint is imposed to the bases. Therefore, reasonable initial values of $\bar{\mathbf{B}}_0$ and Φ_0 may speed up convergence and achieve better results, though the RCPF algorithm is not sensitive to initialization. In our applications, we adopt a rigid approximation of the object and utilize the weak perspective rank-3 factorization method [13, 18, 19] to recover the average rigid structure $\bar{\mathbf{S}}_r \in \mathbb{R}^{3\times n}$ and motion $\mathbf{R}_{wi}$ of the sequence. The initial values are constructed as

$$\Phi_0 = \begin{bmatrix} \mathbf{R}_{w1} & \cdots & \mathbf{R}_{w1} \\ \vdots & \ddots & \vdots \\ \mathbf{R}_{wm} & \cdots & \mathbf{R}_{wm} \end{bmatrix}, \qquad \bar{\mathbf{B}}_0 = \begin{bmatrix} \bar{\mathbf{S}}_r + \mathbf{N}_1 \\ \vdots \\ \bar{\mathbf{S}}_r + \mathbf{N}_k \end{bmatrix} \tag{7.18}$$

where $\mathbf{N}_i \in \mathbb{R}^{3\times n}$ is a small random matrix which ensures that the initial shape bases are independent of each other. The initialization are based on two observations: (1) Most nonrigid objects are rigid dominate and may be approximated by a rigid assumption. The motions recovered by rigid factorization are usually close to the real solution, therefore we update the rotation matrix at the last step of the algorithm, though the order of these steps can be interchanged. (2) All shape bases of the same nonrigid object have almost the same size in 3D space. Thus they are initialized as the mean shape with random noise.

The proposed RCPF algorithm is similar to the iterative optimization scheme proposed by Torresani *et al.* [23], where the authors start with a rigid approximation and then optimize the shape bases, deformation weight and rotation matrix iteratively by a tri-linear algorithm. However, the initial deformation weights are chosen randomly in [23], thus it may not guarantee the recovery of a uniformly distributed shape bases. One may recover a small/large basis with a large/small deformation weight. While the initial shape bases constructed in (7.18) can avoid the unstable situation. The rotation matrix in [23] is parameterized by exponential coordinates with 3 variables, thus the orthonormality of the rotation matrix is preserved. However, the linearization is just an approximation of (7.9). Experiment shows that the proposed updating method in (7.14) and (7.15) usually achieves better result.

7.3.3 Sequential Factorization

We have now recovered the motions, shape bases and deformation weights of m frames in the sequence. If these frames embody the object deformation, then the shape and motion matrices of any new frame can be quickly fitted from the pre-learned models. The model fitting problem is well studied in the literature, such as the active appearance model (AAM) [9] for 2D cases, the morphable model (MM) [1] for 3D cases and their combination for 2D+3D cases [29]. Both AAM and MM learn a set of model parameters that control the modes of shape and appearance, and then generate the new shape/appearance as a base shape/appearance plus a linear combination of some shape/appearance matrices. Inspired by these ideas, we propose a sequential factorization method to model the structure and motion of new frames.

Suppose

$$\bar{\mathbf{B}} = \begin{bmatrix} \mathbf{B}_1 \\ \cdots \\ \mathbf{B}_k \end{bmatrix}$$

is the pre-learned shape bases, $\mathbf{W}_x \in \mathbb{R}^{2\times n}$ is the tracking matrix of the new frame, then from

$$\mathbf{W}_x = \mathbf{M}_x\bar{\mathbf{B}} = [\omega_{x1}\mathbf{R}_{Ax}, \ldots, \omega_{xk}\mathbf{R}_{Ax}]\bar{\mathbf{B}} \tag{7.19}$$

we have

$$\mathbf{M}_x = [\omega_{x1}\mathbf{R}_{Ax}, \ldots, \omega_{xk}\mathbf{R}_{Ax}] = \mathbf{W}_x\bar{\mathbf{B}}^T\left(\bar{\mathbf{B}}\bar{\mathbf{B}}^T\right)^{-1} \tag{7.20}$$

Thus the rotation matrix $\mathbf{R}_{Ax}$ and the deformation weights ω_{xl} can be directly factorized from $\mathbf{M}_x$ via Procrustes analysis [4, 23]. The solution may be further optimized iteratively in a similar way as in step 3 of the RCPF algorithm. The corresponding structure can be obtained from

$$\mathbf{S}_x = \sum_{l=1}^{k}\omega_{xl}\mathbf{B}_l \tag{7.21}$$

This is a simplified model fitting problem. The sequential factorization algorithm is actually a direct extension of the RCPF algorithm to single frame cases. It utilizes the pre-learned shape bases and only handles the tracking matrix of a specific frame, thus the algorithm is very fast and workable in real time.

7.4 Evaluations on Synthetic Data

During simulations, we generated a $20 \times 20 \times 20$ cube with three visible surfaces in space. There are 20 evenly distributed points on each edge and three sets of moving points (21×3 points) on the surfaces of the cube that move at constant speed on the surfaces as shown in Fig. 7.1. We generated 20 cubes (together with the three moving parts) in space with randomly selected poses, then projected each cube to an image by perspective projection. The image size is 800×800 and there are 243 image points in each frame (180 points belong to the cube and the rest 63 points belong to the three moving parts). The distance from the camera to the object was set at about 240, which is very close to weak perspective assumption.

7.4.1 Reconstruction Results and Evaluations

The simulated object in the test can be taken as a nonrigid object that deforms at a constant speed. We recovered the structure and motion of the synthetic sequence by the proposed RCPF algorithm and compared it with two related methods. The first one is a SVD-based algorithm with basis constraint [28]; the second one is the tri-linear optimization algorithm of [23]. Figure 7.1 shows the recovered 3D structures of two frames. The results are visually similar and very close to the ground truths.

The reconstructed structure in Fig. 7.1 is defined up to a 3D similarity transformation with the ground truth. For convenience of evaluation, we calculate the transformation matrix from the point correspondences between the recovered structure and its ground truth, then register each reconstructed shape with its associated ground truth. In order to give a statistically meaningful comparison, we add Gaussian noise to the tracking matrix and vary the noise level from 0 to 3 pixels in steps of 0.5. At each noise level, we take 100 independent tests to recover the structure by the three algorithms and compute the reconstruction error as the point-to-point distance between the registered structure and its ground truth. The mean and standard deviation at each noise level are shown in Fig. 7.2. We can see from the result that the proposed algorithm outperforms both the tri-linear and the SVD-based methods. We also tested sequential factorization algorithm for some new frames, the reconstruction errors are comparable to the RCPF algorithm for the entire sequence (results are not included).

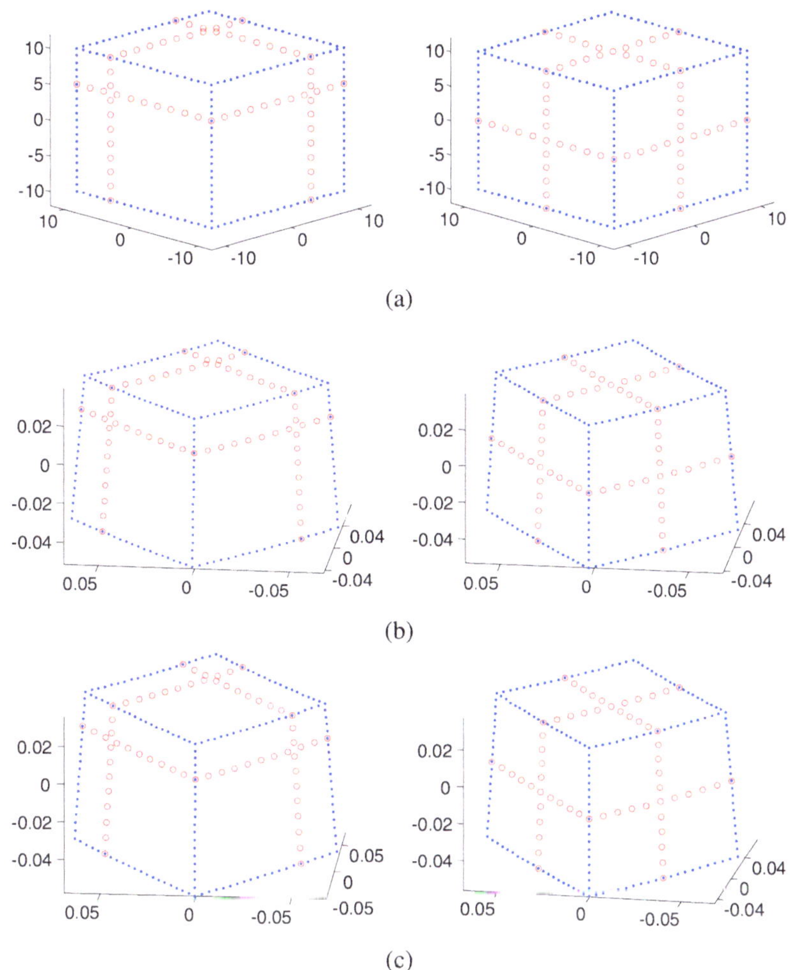

Fig. 7.1 Synthetic data and reconstruction results by different algorithms. (**a**) Two generated structures of the synthetic cube (*in blue dot*) and the three sets of moving points (*in red circle*); (**b**) The recovered structures of the two frames by the RCPF algorithm; (**c**) The corresponding structures recovered by SVD-based algorithm

7.4.2 Convergence Property

We tested convergence of the algorithm in three conditions: First, we implemented the RCPF algorithm to recover shape and motion from 20 frames without added noise. Second, we randomly deleted 20% entries from the initial tracking matrix and performed the RCPF using the remaining data. In the third situation, we added 2-pixel Gaussian noise to the original tracking matrix. At each iteration, we recorded the variation of the reprojected

7

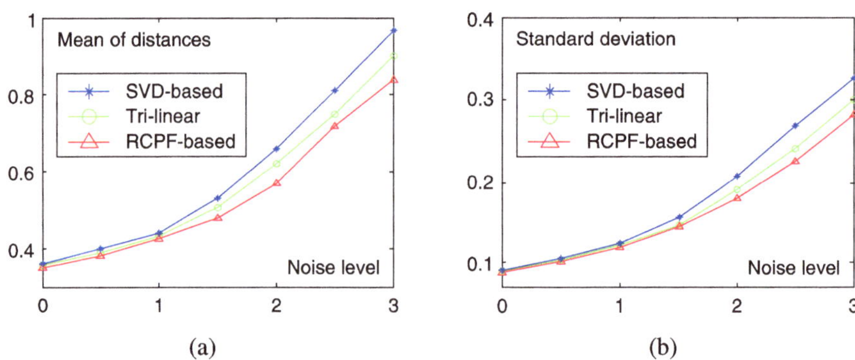

Fig. 7.2 Evaluation and comparison between different algorithms. (**a**) The mean of the reconstruction errors at different noise level; (**b**) The corresponding standard deviations

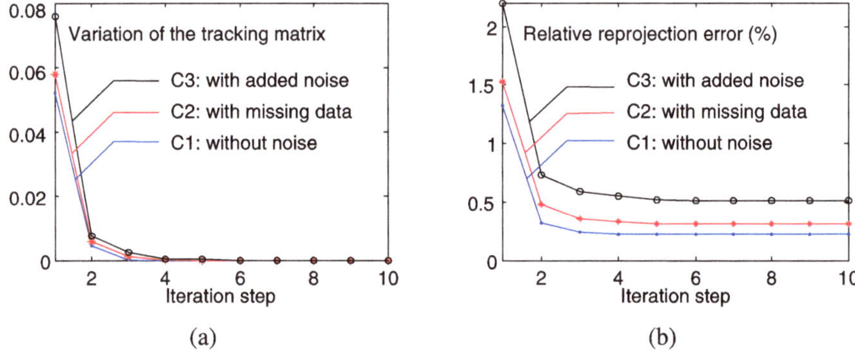

Fig. 7.3 The convergence property of RCPF algorithm. (**a**) The variation of the reprojected tracking matrix at each iteration; (**b**) The relative reprojection error of the algorithm at each iteration

tracking matrix (7.17) and the relative reprojection error which is defined as

$$\mathbf{E}_{rep} = \frac{\|\mathbf{W} - \mathbf{W}_t\|_F}{\|\mathbf{W}\|_F} \times 100\ (\%) \tag{7.22}$$

where $\mathbf{W}_t$ is the reprojected tracking matrix at the tth iteration. As evident from Fig. 7.3, the RCPF algorithm converges quickly, even with some missing data and measurement errors. It seems that the missing data does not have noticeable influence on the convergence speed. However, the algorithm may fail when the proportion of missing entries increases to a certain level as noticed in [13]. One may also find that a small residual error still exists even without noise, this is because the images are generated by perspective projection rather than affine projection. The residual error will also increase with an increase in missing data and noise level.

Table 7.1 Performance comparison of different algorithms with respect to the relative distances from object to the cameras

Relative distance		7	8	9	10	11	12	13	14
	RCPF-based	5.306	4.848	4.573	4.317	4.076	3.813	3.676	3.396
$E_{rep}(\%)$	Tri-linear	5.482	4.984	4.802	4.497	4.126	3.911	3.755	3.444
	SVD-based	5.672	5.265	4.973	4.622	4.284	4.087	3.833	3.522
Iteration	RCPF-based	7.7	7.2	6.5	5.9	5.4	5.1	5.0	5.0
	Tri-linear	8.2	7.8	7.1	6.7	6.1	5.4	5.0	5.0

7.4.3 Influence of Imaging Conditions

The factorization algorithm is based on affine assumption. We studied the influence of different imaging conditions on the algorithm's performance. We vary the relative distance (i.e. the ratio of the distance from the object to the camera over the object depth) from 7 to 14 and generated different sets of images. At each position, we recover the structure of each frame and calculate the relative perspective reprojection errors by different algorithms. The results are tabulated in Table 7.1, where the errors are evaluated by 100 independent tests. We also record the iteration times of the RCPF and the tri-linear algorithms. We see from the results that the proposed RCPF algorithm performs better than the other two methods. The reconstruction error increases as the camera moves close to the object, but good solutions are usually guaranteed when the relative distance is greater than 8. However, all methods may fail when the affine condition is not satisfied.

7.5 Evaluations on Real Sequences

We tested the proposed methods on several image sequences and we report the results obtained using three sequences.

7.5.1 Test on Grid Sequence

The sequence was captured by a Canon Powershot G3 camera. There are 12 images with a resolution of 1024×768. On the two orthogonal background surfaces, there are three moving objects that move linearly in three directions. Figure 7.4 shows two frames of the sequence with 206 tracked features across the sequence, where 140 features belong to the static background and 66 features belong to the three moving objects. We recovered the 3D structure of the scene by the RCPF algorithm. Figure 7.4 shows the reconstructed VRML

7

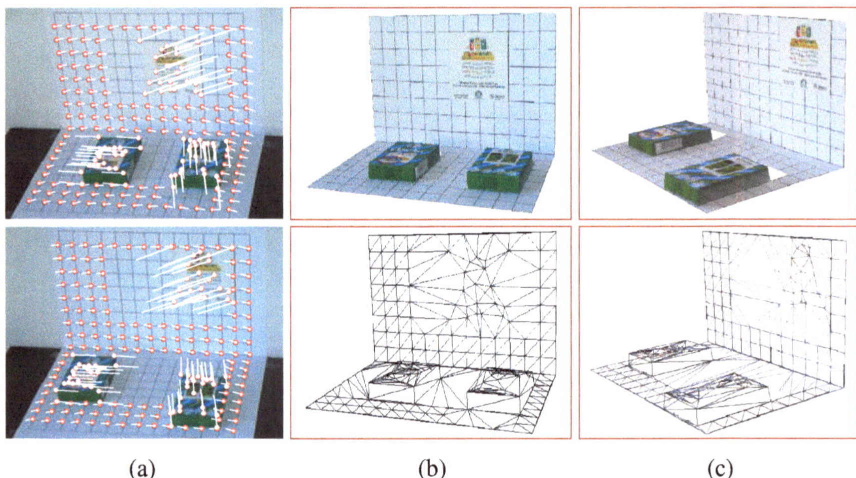

(a) (b) (c)

Fig. 7.4 Reconstruction result of grid sequence. (**a**) Two frames from the sequence overlaid with 206 tracked features and the relative disparities shown in *white lines*; (**b**) & (**c**) The reconstructed VRML models and corresponding wireframes shown from different viewpoints

Table 7.2 Performance comparison and evaluation of different algorithms for real sequences

Algorithm	$\mathbf{E}_{angb}$ (rad)	$\mathbf{E}_{ratio}$ (rad)	$\mathbf{E}_{angd}$ (rad)	$\mathbf{E}_{rep1}$ (%)	$\mathbf{E}_{rep2}$ (%)	$\mathbf{E}_{rep3}$ (%)
RCPF-based	0.030	0.128	0.014	4.584	6.390	4.952
Tri-linear	0.035	0.132	0.014	4.927	6.407	4.996
SVD-based	0.041	0.147	0.016	5.662	6.425	5.058

models and the corresponding triangulated wireframes of from different viewpoints. The dynamic structure of the scene is correctly recovered by the algorithm.

The background of this sequence consists of two mutually orthogonal sheets with square grids. We take this as a ground truth and recover the angle between the two reconstructed surfaces of the background, the length ratio of the two diagonals of each square and the angle formed by the two diagonals. The mean errors of these three values are denoted by $\mathbf{E}_{angb}$, $\mathbf{E}_{ratio}$, and $\mathbf{E}_{angd}$, respectively. We also calculate the relative reprojection error $\mathbf{E}_{rep1}$. The comparative results obtained by the three algorithms are tabulated in Table 7.2. Results show that the proposed method performs better than the other two.

7.5.2 Test on Franck Sequence

The sequence and tracking data are similar to that in Chap. 6. There are 60 frames with 68 tracked features across the sequence. Figure 7.5 shows the reconstructed VRML models and the triangulated wireframes of two frames by the proposed method. Different facial

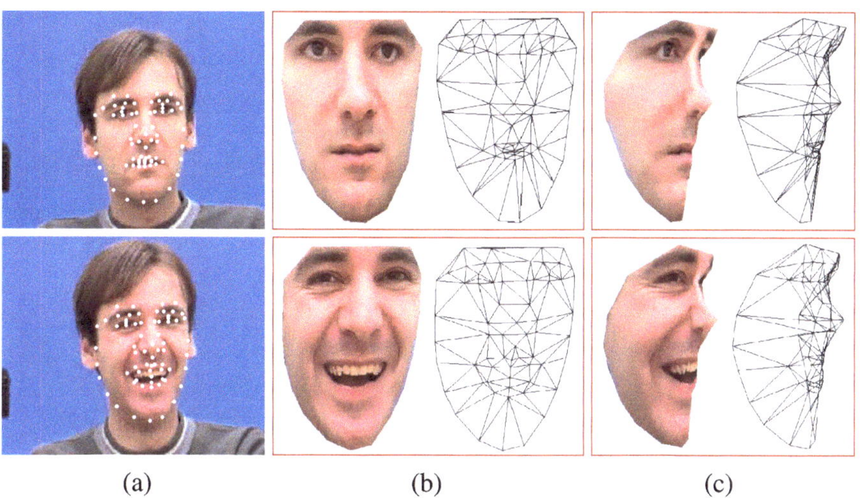

(a) (b) (c)

Fig. 7.5 Reconstruction of different facial expressions from Franck sequence. (**a**) Two frames of the sequence overlaid with the tracked features; (**b**) & (**c**) The front and side views of the reconstructed VRML models and the corresponding wireframes

expressions are correctly recovered as shown in Fig. 7.5. As a comparison, the relative reprojection errors $\mathbf{E}_{rep2}$ by different algorithms are listed in Table 7.2.

7.5.3 Test on Quilt Sequence

The sequence has 28 frames with a resolution of 1024 × 768. The quilt was pressed to deform its shape in the sequence. The correspondences were established by the method in [30] and 1288 features were tracked across the sequence. The reconstructed VRML models and wireframes of three frames are shown in Fig. 7.6. The recovered 3D structure of the quilt with deformation is visually plausible and realistic. The relative reprojection errors $\mathbf{E}_{rep3}$ using the three algorithms are listed in Table 7.2, it is clear that the proposed method gives the best performance.

7.6 Closure Remarks

7.6.1 Conclusion

In this chapter, under the assumption of affine camera, we introduced a rotation constrained power factorization algorithm to recover the shape and motion of nonrigid objects from

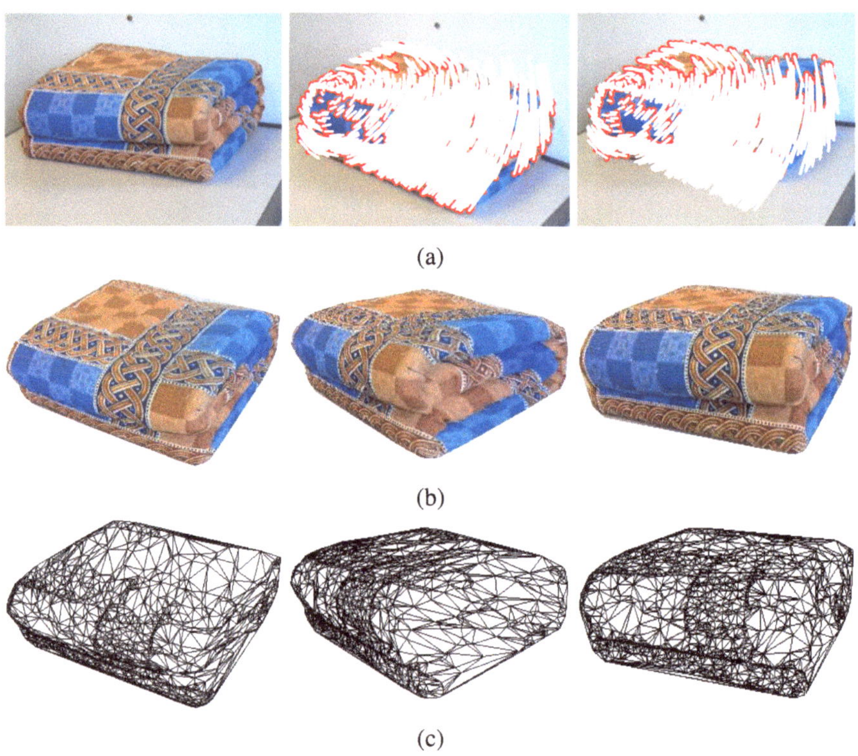

Fig. 7.6 Reconstruction result of quilt sequence. (**a**) Three frames from the sequence with tracked features overlaid on two images; (**b**) Three reconstructed VRML models with textures mapping; (**c**) The corresponding wireframes of the recovered structures

monocular image sequences. The algorithm directly integrates the orthonormal constraint of the rotation matrix and the replicated block structure of the motion matrix into the iterations, thus it avoids the difficult problem of recovering the transformation matrix in the SVD-based method. It is easy to implement and can work with missing data in the tracking matrix. Based on the recovered shape bases of the batch factorization, we also proposed a fast and practical sequential factorization algorithm to recover structure and motion of new frames. The solution may be further optimized via bundle adjustment.

7.6.2 Review Questions

1. *Power factorization.* Explain the process of power factorization. Show how a tracking matrix with incomplete data is factorized.

2. *RCPF algorithm.* Review computation details of the rotation constrained power factorization algorithm. How to ensure the orthonormality and block replicated structure of the motion matrix? Can you think of any other initialization method for the algorithm?
3. *Sequential factorization.* How do you factorize the deformation weights and rotation matrix from the recovered motion matrix of a new frame?

References

1. Blanz, V., Vetter, T.: A morphable model for the synthesis of 3D faces. In: Proc. of SIGGRAPH, pp. 187–194 (1999)
2. Brand, M.: Morphable 3D models from video. In: Proc. of IEEE Conference on Computer Vision and Pattern Recognition, vol. 2, pp. 456–463 (2001)
3. Brand, M.: Incremental singular value decomposition of uncertain data with missing values. In: Proc. of European Conference on Computer Vision, pp. 707–721 (2002)
4. Bregler, C., Hertzmann, A., Biermann, H.: Recovering non-rigid 3D shape from image streams. In: Proc. of IEEE Conference on Computer Vision and Pattern Recognition, vol. 2, pp. 690–696 (2000)
5. Buchanan, A.M., Fitzgibbon, A.W.: Damped Newton algorithms for matrix factorization with missing data. In: Proc. of IEEE Conference on Computer Vision and Pattern Recognition, vol. 2, pp. 316–322 (2005)
6. Chen, P.: Optimization algorithms on subspaces: Revisiting missing data problem in low-rank matrix. Int. J. Comput. Vis. **80**(1), 125–142 (2008)
7. Chen, P., Suter, D.: Recovering the missing components in a large noisy low-rank matrix: Application to SFM. IEEE Trans. Pattern Anal. Mach. Intell. **26**(8), 1051–1063 (2004)
8. Christy, S., Horaud, R.: Euclidean shape and motion from multiple perspective views by affine iterations. IEEE Trans. Pattern Anal. Mach. Intell. **18**(11), 1098–1104 (1996)
9. Cootes, T.F., Edwards, G.J., Taylor, C.J.: Active appearance models. IEEE Trans. Pattern Anal. Mach. Intell. **23**(6), 681–685 (2001)
10. Golub, G., Loan, C.V.: Matrix Computations. John Hopkins University Press, Baltimore (1983)
11. Gruber, A., Weiss, Y.: Factorization with uncertainty and missing data: Exploiting temporal coherence. In: Advances in Neural Information Processing Systems (2003)
12. Gruber, A., Weiss, Y.: Multibody factorization with uncertainty and missing data using the em algorithm. In: Proc. of IEEE Conference on Computer Vision and Pattern Recognition, vol. 1, pp. 707–714 (2004)
13. Hartley, R., Schaffalizky, F.: Powerfactorization: 3D reconstruction with missing or uncertain data. In: Proc. of Australia-Japan Advanced Workshop on Computer Vision (2003)
14. Hung, Y., Tang, W.: Projective reconstruction from multiple views with minimization of 2D reprojection error. Int. J. Comput. Vis. **66**(3), 305–317 (2006)
15. Jacobs, D.: Linear fitting with missing data for structure-from-motion. Comput. Vis. Image Underst. **82**(1), 57–81 (2001)
16. Martinec, D., Pajdla, T.: Structure from many perspective images with occlusion. In: Proc. of European Conference on Computer Vision, pp. 355–369 (2002)
17. Morita, T., Kanade, T.: A sequential factorization method for recovering shape and motion from image streams. IEEE Trans. Pattern Anal. Mach. Intell. **19**(8), 858–867 (1997)
18. Poelman, C., Kanade, T.: A paraperspective factorization method for shape and motion recovery. IEEE Trans. Pattern Anal. Mach. Intell. **19**(3), 206–218 (1997)
19. Quan, L.: Self-calibration of an affine camera from multiple views. Int. J. Comput. Vis. **19**(1), 93–105 (1996)

20. Sturm, P.F., Triggs, B.: A factorization based algorithm for multi-image projective structure and motion. In: Proc. of European Conference on Computer Vision, vol. 2, pp. 709–720 (1996)
21. Tomasi, C., Kanade, T.: Shape and motion from image streams under orthography: A factorization method. Int. J. Comput. Vis. **9**(2), 137–154 (1992)
22. Torresani, L., Hertzmann, A., Bregler, C.: Nonrigid structure-from-motion: Estimating shape and motion with hierarchical priors. IEEE Trans. Pattern Anal. Mach. Intell. **30**(5), 878–892 (2008)
23. Torresani, L., Yang, D.B., Alexander, E.J., Bregler, C.: Tracking and modeling non-rigid objects with rank constraints. In: Proc. of IEEE Conference on Computer Vision and Pattern Recognition, vol. 1, pp. 493–500 (2001)
24. Vidal, R., Hartley, R.I.: Motion segmentation with missing data using powerfactorization and GPCA. In: Proc. of IEEE Conference on Computer Vision and Pattern Recognition, vol. 2, pp. 310–316 (2004)
25. Wang, G.: A hybrid system for feature matching based on SIFT and epipolar constraints. Tech. Rep. Department of ECE, University of Windsor (2006)
26. Wang, G., Tsui, H.T., Wu, J.: Rotation constrained power factorization for structure from motion of nonrigid objects. Pattern Recogn. Lett. **29**(1), 72–80 (2008)
27. Wang, G., Wu, J.: Stratification approach for 3-D Euclidean reconstruction of nonrigid objects from uncalibrated image sequences. IEEE Trans. Syst. Man Cybern., Part B **38**(1), 90–101 (2008)
28. Xiao, J., Chai, J.X., Kanade, T.: A closed-form solution to non-rigid shape and motion recovery. In: Proc. of European Conference on Computer Vision, vol. 4, pp. 573–587 (2004)
29. Xiao, J., Kanade, T.: Non-rigid shape and motion recovery: Degenerate deformations. In: Proc. of IEEE Conference on Computer Vision and Pattern Recognition, vol. 1, pp. 668–675 (2004)
30. Yao, J., Cham, W.: Feature matching and scene reconstruction from multiple widely separated views. Tech. Rep., Chinese University of Hong Kong (2005)

Stratified Euclidean Reconstruction 8

Abstract The chapter proposes a stratification approach to recover the structure of nonrigid objects under the assumption that the object is composed of separable rigid features and deformed ones. First, we propose a deformation weight constraint for the problem and prove the invariability between the recovered structure and shape bases under this constraint. Second, we propose a constrained power factorization (CPF) algorithm to recover the deformation structure in affine space. Third, we propose to segment the rigid features from the deformed ones in 3D affine space which makes segmentation more accurate and robust. Finally, we recover the stratification matrix from the rigid features and upgrade the structure from affine to the Euclidean space.

What remains to be resolved is the question of knowing to what extent and up to what point these hypotheses are found to be confirmed by experience.

Bernhard Riemann (1826–1866)

8.1 Introduction

Bregler and his colleagues [3] first proposed to model the problem of nonrigid structure from motion using basis representation, since then great progress has been made on nonrigid factorization [2, 5, 13, 18]. The method was extended to perspective projection model by Xiao and Kanade [19] and Wang *et al.* [15]. As an alternative approach, Wang *et al.* [16] extended the power factorization algorithm [8] to nonrigid factorization. Torresani *et al.* [12] proposed an algorithm to learn the time-varying shape using expectation maximization, where the structure was modeled as a rigid component combined with a nonrigid one. Del Bue *et al.* proposed to segment the rigid part of the object directly from the tracking matrix either from a rank-3 constraint [4] or epipolar constraint [5]. They then recover the nonrigid shape by a constrained nonlinear optimization process. However, the nonlinear process heavily relies on the initial estimates, and the segmentation process may be difficult in cases where different groups of features satisfy the rank constraint [4] or the

G. Wang, Q.M.J. Wu, *Guide to Three Dimensional Structure and Motion Factorization*, Advances in Pattern Recognition,
DOI 10.1007/978-0-85729-046-5_8,

8

inter-frame movements are small [5]. Wang and Wu [17] proposed a more reliable method to segment the rigid and nonrigid features in 3D affine space.

Previous methods utilize the orthonormal constraints [1, 14] and the basis constraints [18] for nonrigid factorization, they do not impose any constraint on the deformation weight. In this chapter, we try to solve the problem from a new viewpoint based on uncalibrated affine camera model [17]. We assume that some part of the nonrigid object is rigid and does not deform across the sequence. Our main idea is to recover the affine structure of the object and segment the rigid features from the deformed ones. Then, we estimate the transformation from affine to metric space from the rigid features and stratify the affine structure to the Euclidean space. We propose a deformation weight constraint for nonrigid factorization and prove that the recovered structure and shape bases are transformation invariant under this constraint. We also propose a constrained power factorization algorithm to factorize the tracking matrix in affine space.

The remaining chapter is organized as follows: The deformation weight constraint is analyzed in Sect. 8.2. The constrained power factorization algorithm is elaborated in Sect. 8.3 followed by the strategies for deformation detection and Euclidean stratification in Sect. 8.4. Finally, some experimental evaluations on synthetic and real sequences are given in Sects. 8.5 and 8.6 respectively.

8.2 Deformation Weight Constraint

In this section, we first give a brief review of nonrigid factorization. Then, we propose a new constraint to the deformation weights and provide a detailed analysis on the constraint.

8.2.1 Nonrigid Factorization

Following Bregler's assumption [3], the nonrigid shape in Euclidean space is approximated by a linear combination of k shape bases as

$$\bar{\mathbf{S}}_i = \sum_{l=1}^{k} \omega_{il} \mathbf{B}_l \tag{8.1}$$

where $\mathbf{B}_l \in \mathbb{R}^{3\times n}$ are the shape bases that embody the principal modes of deformation; $\omega_{il} \in \mathbb{R}$ are the deformation weights.

Under affine camera model, the projection from a space point $\bar{\mathbf{X}}_j$ to an image point $\bar{\mathbf{x}}_{ij}$ can be expressed as $\bar{\mathbf{x}}_{ij} = \mathbf{A}_i \bar{\mathbf{X}}_j$ if we register all points to the centroid, where $\mathbf{A}_i$ is a 2×3 affine projection matrix. Given a tracking matrix $\mathbf{W} \in \mathbb{R}^{2m\times n}$ of n points across m frames of a nonrigid object, the nonrigid factorization equation can be written as

$$\mathbf{W}_{2m\times n} = \mathbf{M}_{2m\times 3k} \bar{\mathbf{B}}_{3k\times n} \tag{8.2}$$

where

$$\mathbf{M} = \begin{bmatrix} \omega_{11}\mathbf{A}_1 & \cdots & \omega_{1k}\mathbf{A}_1 \\ \vdots & \ddots & \vdots \\ \omega_{m1}\mathbf{A}_m & \cdots & \omega_{mk}\mathbf{A}_m \end{bmatrix}, \qquad \bar{\mathbf{B}} = \begin{bmatrix} \mathbf{B}_1 \\ \vdots \\ \mathbf{B}_k \end{bmatrix} \tag{8.3}$$

The factorization can be carried out by SVD decomposition with rank-$3k$ constraint. The decomposition is defined up to a nonsingular $3k \times 3k$ upgrading matrix $\mathbf{H}$ which is normally recovered via rotation and basis constraints [18]. Nevertheless, the nonrigid factorization algorithm does not work as perfect as the rigid case due to complications in recovering the upgrading matrix and the block replicated motion matrix in (8.3).

8.2.2 Deformation Weight Constraint

Previous studies on nonrigid factorization do not impose any constraints on deformation weights. We will define a deformation constraint here and show some useful properties of the constraint.

Definition 8.1 (Deformation Weight Constraint) In nonrigid factorization (8.2), the recovered deformation weights corresponding to each frame satisfy the constraint of unit summation.

$$\sum_{l=1}^{k} \omega_{il} = 1, \quad \forall i = 1, \ldots, m \tag{8.4}$$

Under the deformation weight constraint, we have the following properties.

Theorem 8.1 *When the structure $\bar{\mathbf{S}}_i$ of each frame and the shape bases $\mathbf{B}_l$ are transformed by an arbitrary Euclidean transformation, the relationship in* (8.1) *remains invariant if and only if the deformation weight constraint is satisfied.*

Proof The transformation in Euclidean space can be termed as either a Euclidean transformation

$$\mathbf{H}_e = \begin{bmatrix} \mathbf{R} & \mathbf{t} \\ \mathbf{0}^T & 1 \end{bmatrix} \tag{8.5}$$

or a similarity transformation

$$\mathbf{H}_s = \begin{bmatrix} s\mathbf{R} & \mathbf{t} \\ \mathbf{0}^T & 1 \end{bmatrix} \tag{8.6}$$

where $\mathbf{R}$ is a 3×3 orthonormal rotation matrix, $\mathbf{t}$ is a 3D translation vector, $\mathbf{0}$ is a null 3-vector, and s is a similarity scalar. Without loss of generality, let us take $\mathbf{H}_s$ as an example.

Suppose $\mathbf{B}_l$ and ω_{il} are one set of Euclidean shape bases and deformation weights associated with the ith frame. The corresponding structure of the frame is $\bar{\mathbf{S}}_i = \sum_{l=1}^{k} \omega_{il} \mathbf{B}_l$ as given by (8.1). Under the similarity transformation (8.6), the shape bases and structure are transformed to $\mathbf{B}'_l$ and $\bar{\mathbf{S}}'_i$ as

$$\mathbf{B}'_l = s\mathbf{R}\mathbf{B}_l + \mathbf{T} \tag{8.7}$$

$$\bar{\mathbf{S}}'_i = s\mathbf{R}\bar{\mathbf{S}}_i + \mathbf{T} \tag{8.8}$$

where $\mathbf{T} = [\mathbf{t}, \mathbf{t}, \ldots, \mathbf{t}]$ is a $3 \times n$ matrix. From (8.7) and (8.8), we have

$$\begin{aligned} \bar{\mathbf{S}}'_i &= s\mathbf{R}\left(\sum_{l=1}^{k} \omega_{il} \mathbf{B}_l\right) + \mathbf{T} = \sum_{l=1}^{k} \omega_{il} \left(s\mathbf{R}\mathbf{B}_l\right) + \mathbf{T} \\ &= \sum_{l=1}^{k} \omega_{il} \mathbf{B}'_l + \left(1 - \sum_{l=1}^{k} \omega_{il}\right)\mathbf{T} \end{aligned} \tag{8.9}$$

It is clear that the transformed structure and shape bases satisfy the relation (8.1) as

$$\bar{\mathbf{S}}'_i = \sum_{l=1}^{k} \omega_{il} \mathbf{B}'_l \tag{8.10}$$

if and only if the constraint (8.4) is satisfied. □

Theorem 8.1 tells us that if the recovered deformation weights satisfy the constraint (8.4), the relationship between the object structure and its shape bases is invariant under any transformation in Euclidean space.

Theorem 8.2 *Suppose the homogeneous forms of* $\bar{\mathbf{S}}_i$ *and* $\mathbf{B}_l$ *are*

$$\tilde{\mathbf{S}}_i = \begin{bmatrix} \bar{\mathbf{S}}_i \\ \mathbf{1}^T \end{bmatrix}, \qquad \tilde{\mathbf{B}}_l = \begin{bmatrix} \mathbf{B}_l \\ \mathbf{1}^T \end{bmatrix} \tag{8.11}$$

where $\mathbf{1}$ *stands for a n-vector with unit entities. Then* (8.1) *can be written in homogeneous form as*

$$\tilde{\mathbf{S}}_i = \sum_{l=1}^{k} \omega_{il} \tilde{\mathbf{B}}_l \tag{8.12}$$

if and only if the deformation weight constraint (8.4) *is satisfied.*

Proof The result is obvious. From

$$\sum_{l=1}^{k} \omega_{il} \tilde{\mathbf{B}}_l = \sum_{l=1}^{k} \omega_{il} \begin{bmatrix} \mathbf{B}_l \\ \mathbf{1}^T \end{bmatrix} = \begin{bmatrix} \sum_{l=1}^{k} \omega_{il} \mathbf{B}_l \\ \sum_{l=1}^{k} \omega_{il} \mathbf{1}^T \end{bmatrix} \tag{8.13}$$

we have

$$\sum_{l=1}^{k} \omega_{il} \tilde{\mathbf{B}}_l = \begin{bmatrix} \bar{\mathbf{S}}_i \\ \mathbf{1}^T \end{bmatrix} = \tilde{\mathbf{S}}_i \tag{8.14}$$

if and only if the deformation weight constraint is satisfied. □

Theorem 8.3 *When the structure and the shape bases are transformed by an arbitrary affine transformation, the relationship* (8.1) *remains invariant if and only if the deformation weight constraint is also satisfied.*

Proof The transformation from Euclidean space to affine space can be modeled by an affine transformation

$$\mathbf{H}_a = \begin{bmatrix} \mathbf{P} & \mathbf{t} \\ \mathbf{0}^T & 1 \end{bmatrix} \tag{8.15}$$

where $\mathbf{P}$ is an invertible 3×3 matrix, and $\mathbf{t}$ is a 3D translation vector. We will use the result of Theorem 8.2 to prove the theorem. Let $\mathbf{B}'_l$ and $\bar{\mathbf{S}}'_i$ be the affine transformed shape bases and structure of $\mathbf{B}_l$ and $\bar{\mathbf{S}}_i$ respectively. Their relationship in homogeneous form is

$$\tilde{\mathbf{B}}'_l = \mathbf{H}_a \tilde{\mathbf{B}}_l, \qquad \tilde{\mathbf{S}}'_i = \mathbf{H}_a \tilde{\mathbf{S}}_i \tag{8.16}$$

From theorem 8.2, we have

$$\tilde{\mathbf{S}}'_i = \mathbf{H}_a \sum_{l=1}^{k} \omega_{il} \tilde{\mathbf{B}}_l = \sum_{l=1}^{k} \omega_{il} \mathbf{H}_a \tilde{\mathbf{B}}_l = \sum_{l=1}^{k} \omega_{il} \tilde{\mathbf{B}}'_l \tag{8.17}$$

It is clear that equation (8.17) can be written in the following inhomogeneous form

$$\bar{\mathbf{S}}'_l - \sum_{l=1}^{k} \omega_{il} \mathbf{B}'_l \tag{8.18}$$

if and only if the deformation weight constraint is satisfied. □

8.2.3 Geometrical Explanation

During nonrigid factorization, we register all image measurements to the centroid. Thus, the recovered shape bases, as well as the structure of each frame, are also registered to their corresponding centroids. When the shape bases and the structure are subjected to a Euclidean/affine transformation, their centroids are deviated by a translation vector. With the deformation constraint, it is guaranteed that the translation term that resulted from the combination of shape bases is consistent with that of the transformed structure, such that they are invariant to the transformation. The following example illustrates the effect of deformation weight constraint.

Given a tracking matrix $\mathbf{W}$, suppose there are three frames and two shape bases. In the first case, we obtain the following factorization.

$$\mathbf{W} = \mathbf{MB} = \begin{bmatrix} 0.2\mathbf{A}_1 & 0.8\mathbf{A}_1 \\ 0.4\mathbf{A}_2 & 0.6\mathbf{A}_2 \\ 0.6\mathbf{A}_3 & 0.4\mathbf{A}_3 \end{bmatrix} \begin{bmatrix} \mathbf{B}_{11} \\ \mathbf{B}_{12} \end{bmatrix} \tag{8.19}$$

where the deformation weights satisfy the constraint (8.4). Since the decomposition is not unique and we can insert any invertible transformation matrix $\mathbf{H}$ into (8.19). Suppose the transformation is

$$\mathbf{H} = \begin{bmatrix} \mathbf{I}_3 & \mathbf{0} \\ \mathbf{0} & 0.5\mathbf{I}_3 \end{bmatrix} \tag{8.20}$$

where $\mathbf{I}_3$ is a 3×3 identity matrix. Then the factorization (8.19) can be written as

$$\mathbf{W} = (\mathbf{MH})(\mathbf{H}^{-1}\mathbf{B}) = \begin{bmatrix} 0.2\mathbf{A}_1 & 0.4\mathbf{A}_1 \\ 0.4\mathbf{A}_2 & 0.3\mathbf{A}_2 \\ 0.6\mathbf{A}_3 & 0.2\mathbf{A}_3 \end{bmatrix} \begin{bmatrix} \mathbf{B}_{21} \\ \mathbf{B}_{22} \end{bmatrix} \tag{8.21}$$

where the transformed new shape bases $\mathbf{B}_{21} = \mathbf{B}_{11}$ and $\mathbf{B}_{22} = 2\mathbf{B}_{12}$. In the new factorization (8.21), the deformation weight constraint is no longer satisfied. Nevertheless, the corresponding structures of each frame before and after transformation $\mathbf{H}$ remain the same.

$$\begin{cases} \bar{\mathbf{S}}_1 = 0.2\mathbf{B}_{11} + 0.8\mathbf{B}_{12} = 0.2\mathbf{B}_{21} + 0.4\mathbf{B}_{22} \\ \bar{\mathbf{S}}_2 = 0.4\mathbf{B}_{11} + 0.6\mathbf{B}_{12} = 0.4\mathbf{B}_{21} + 0.3\mathbf{B}_{22} \\ \bar{\mathbf{S}}_3 = 0.6\mathbf{B}_{11} + 0.4\mathbf{B}_{12} = 0.6\mathbf{B}_{21} + 0.2\mathbf{B}_{22} \end{cases} \tag{8.22}$$

If we impose the Euclidean transformation (8.5) to the shape bases and structures, we have

$$\mathbf{B}'_{11} = \mathbf{RB}_{11} + \mathbf{T}; \qquad \mathbf{B}'_{12} = \mathbf{RB}_{12} + \mathbf{T} \tag{8.23}$$

$$\mathbf{B}'_{21} = \mathbf{RB}_{21} + \mathbf{T}; \qquad \mathbf{B}'_{22} = \mathbf{RB}_{22} + \mathbf{T} \tag{8.24}$$

$$\bar{\mathbf{S}}'_i = \mathbf{R}\bar{\mathbf{S}}_i + \mathbf{T}, \quad \forall i = 1, 2, 3 \tag{8.25}$$

From (8.22) to (8.25) we verify that

$$\begin{cases} \mathbf{S}'_1 = 0.2\mathbf{B}'_{11} + 0.8\mathbf{B}'_{12} \neq 0.2\mathbf{B}'_{21} + 0.4\mathbf{B}'_{22} \\ \mathbf{S}'_2 = 0.4\mathbf{B}'_{11} + 0.6\mathbf{B}'_{12} \neq 0.4\mathbf{B}'_{21} + 0.3\mathbf{B}'_{22} \\ \mathbf{S}'_3 = 0.6\mathbf{B}'_{11} + 0.4\mathbf{B}'_{12} \neq 0.6\mathbf{B}'_{21} + 0.2\mathbf{B}'_{22} \end{cases} \tag{8.26}$$

By comparing (8.22) and (8.26), we conclude that the relationship between the transformed structures and the transformed shape bases remains invariant if the deformation weight constraint is satisfied.

Corollary 8.1 *In affine space, the nonrigid structure can also be written as a linear combination of a set of affine shape bases just as that in the Euclidean case. It is obvious that the affine solution can be upgraded to the Euclidean space and ensures that their combination relationship invariant if the deformation weight constraint is satisfied.*

Corollary 8.1 suggests the feasibility of solving the nonrigid factorization by a stratification approach. In many cases, we have no knowledge about the camera parameters of an uncalibrated image sequence. Thus, it is difficult to directly adopt the orthonormal constraints to compute the upgrading matrix **H** and recover the Euclidean structure. However, as shown in the following section, we first decompose the tracking matrix in affine space and then stratify it to the Euclidean space.

8.3 Affine Structure and Motion Recovery

As discussed in Chap. 7, the general power factorization method was proposed to find a low-rank approximation of a tracking matrix [8]. However, the replicated block structure of the motion matrix in nonrigid factorization is not observed in a general power factorization method. We will introduce a constrained power factorization (CPF) algorithm to solve this problem.

8.3.1 Constrained Power Factorization

Let us decompose the motion matrix (8.3) into two parts as follows.

$$\mathbf{M} = \Omega \otimes \Phi = \begin{bmatrix} \omega_{11}\mathbf{E} & \cdots & \omega_{1k}\mathbf{E} \\ \vdots & \ddots & \vdots \\ \omega_{m1}\mathbf{E} & \cdots & \omega_{mk}\mathbf{E} \end{bmatrix} \otimes \begin{bmatrix} \mathbf{A}_1 & \cdots & \mathbf{A}_1 \\ \vdots & \ddots & \vdots \\ \mathbf{A}_m & \cdots & \mathbf{A}_m \end{bmatrix} \tag{8.27}$$

where Ω and Φ denote the weighting part and the affine motion part respectively; **E** is a 2×3 matrix with unit entries, and '$\otimes$' stands for the Hadamard product of element-by-element multiplication. The algorithm is summarized as follows.

Algorithm: Constrained power factorization. Given a tracking matrix $\mathbf{W} \in \mathbb{R}^{2m \times n}$, initial motion $\Phi_0 \in \mathbb{R}^{2m \times 3k}$, and shape matrix $\bar{\mathbf{B}}_0 \in \mathbb{R}^{3k \times n}$. Repeat the following three steps until the convergence of product $(\Omega_t \otimes \Phi_t)\bar{\mathbf{B}}_t$.

1. Given Φ_{t-1} and $\bar{\mathbf{B}}_{t-1}$, find Ω_t to minimize $\|\mathbf{W} - (\Omega_t \otimes \Phi_{t-1})\bar{\mathbf{B}}_{t-1}\|_F^2$, subject to the condition that Ω_t satisfies the deformation weight constraint;
2. Given Φ_{t-1} and Ω_t, find $\bar{\mathbf{B}}_t$ to minimize $\|\mathbf{W} - (\Omega_t \otimes \Phi_{t-1})\bar{\mathbf{B}}_t\|_F^2$;

8

3. Given Ω_t and $\bar{\mathbf{B}}_t$, find Φ_t to minimize $\|\mathbf{W} - (\Omega_t \otimes \Phi_t)\bar{\mathbf{B}}_t\|_F^2$, subject to the constraint that Φ_t is a block replicated matrix.

In contrast to the rotation constrained power factorization algorithm discussed in Chap. 7, we combine both the weight constraint and the replicated block structure of the motion matrix into the factorization algorithm. It is quite easy to apply the deformation weight constraint in step 1, since we can always set

$$\omega_{ik} = 1 - \sum_{l=1}^{k-1} \omega_{il}$$

The block replicated structure of Φ in step 3 can be applied as follows.

Let us denote $\mathbf{W}_i \in \mathbb{R}^{2\times n}$ as the ith two-row of $\mathbf{W}$, which is the tracking matrix of ith frame. Then we have

$$\mathbf{W}_i = \mathbf{A}_i \bar{\mathbf{S}}_i = \mathbf{A}_i \left(\sum_{l=1}^{k} \omega_{il} \mathbf{B}_l \right) \tag{8.28}$$

where ω_{il} can be obtained from Ω_t in step 1, $\mathbf{B}_l$ can be obtained from $\bar{\mathbf{B}}_t$ in step 2. Therefore, $\mathbf{A}_i$ may be computed by least-squares as

$$\mathbf{A}_i = \mathbf{W}_i \left(\bar{\mathbf{S}}_i\right)^T \left(\bar{\mathbf{S}}_i \bar{\mathbf{S}}_i^T\right)^{-1} \tag{8.29}$$

Then, the block replicated matrix Φ_t can be constructed from $\mathbf{A}_i$ as indicated in expression (8.27).

Each minimization step in the algorithm is equivalent to solving a set of equations by least-squares. Similar to the RCPF algorithm proposed in Chap. 7, the algorithm can work with missing data. We may only use available entries in the tracking matrix to update Ω, $\bar{\mathbf{B}}$, and Φ. Let Ξ denote the set of available data in the tracking matrix. Then the cost function is modified to

$$\sum_{(i,j)\in\Xi} \left| W_{ij} - \left((\Omega \otimes \Phi)\bar{\mathbf{B}}\right)_{ij} \right|^2 \tag{8.30}$$

8.3.2 Initialization and Convergence Determination

Like general power factorization, the algorithm can work with random initial values. However, the solution upon convergence may not be unique since factorization is defined up to an affine transformation. In the worst case, the recovered affine structure may be stretched or squeezed greatly along a certain direction due to bad initialization, which makes it difficult to detect the deformation features in the subsequent step. Usually, a reasonable initial values simultaneously avoids the worst situation and improves convergence speed.

We start with rigid approximation. Suppose the rank-3 factorization of the tracking matrix is $\mathbf{W} = \hat{\mathbf{M}}\hat{\mathbf{B}}$, where $\hat{\mathbf{M}} \in \mathbb{R}^{2m\times 3}$ is a rigid motion matrix, $\hat{\mathbf{B}} \in \mathbb{R}^{3\times n}$ is an average shape of the object. From this estimation, we compute the mean reprojection error of each point across the sequence and denote it as $\mathbf{e}_r$. Then, the initial values may be constructed as

$$\Phi_0 = [\hat{\mathbf{M}}, \dots, \hat{\mathbf{M}}]_{2m\times 3k} \tag{8.31}$$

$$\bar{\mathbf{B}}_0 = \begin{bmatrix} \hat{\mathbf{S}}_1 + \mathbf{e}_r^T \otimes \mathbf{N}_1 \\ \cdots \\ \hat{\mathbf{S}}_1 + \mathbf{e}_r^T \otimes \mathbf{N}_k \end{bmatrix}_{3k\times n} \tag{8.32}$$

where $\mathbf{e}_r^T \otimes \mathbf{N}_i$ is a reprojection-error-weighted shape balance matrix, $\mathbf{N}_i \in \mathbb{R}^{3\times n}$ is a small random matrix. This term is used to ensure that each initial shape bases are independent of each other. Experiments demonstrate that such an initialization obtains good results.

Akin to the RCPF algorithm, the convergence of the algorithm is determined by checking the variation of reprojected tracking matrix. Suppose $\mathbf{W}_t = (\Omega_t \otimes \Phi_t)\mathbf{B}_t$ is the reprojected tracking matrix at tth iteration, the variation is defined as

$$\delta = \|\mathbf{W}_t - \mathbf{W}_{t-1}\|_F^2 \tag{8.33}$$

8.4 Segmentation and Stratification

8.4.1 Deformation Detection Strategy

From the CPF algorithm, we obtain the 3D affine structure $\bar{\mathbf{S}}_i$ associated with each frame. For most nonrigid objects, some part of its structure is usually non-deformable and can be taken as rigid or near-rigid, while the remaining part is deformable. Suppose the features that belong to the rigid and nonrigid parts are $\bar{\mathbf{S}}_{ri} \in \mathbb{R}^{3\times n1}$ and $\bar{\mathbf{S}}_{ni} \in \mathbb{R}^{3\times n2}$ respectively, where $n1$ and $n2$ are the number of non-deformable and deformable features and $n1 + n2 = n$. Our objective is to separate the rigid features from the deformed ones.

The strategy is to register all affine structures in one reference frame. The rigid parts will be aligned with each other if we register all the 3D structures by virtue of the rigid features since the rigid structure does not change across the sequence. Whereas the deformation part deforms with time and can be easily detected from the registration errors.

The registration in 3D affine space is defined by affine transformation. Let us take the first frame as reference. Then, the transformation from ith frame to the first frame is denoted as

$$\mathbf{H}_{ri} = \begin{bmatrix} \mathbf{P}_i & \mathbf{t}_i \\ \mathbf{0}^T & 1 \end{bmatrix} \tag{8.34}$$

which can be simply computed from

$$\tilde{\mathbf{S}}_{r1} = \mathbf{H}_{ri}\tilde{\mathbf{S}}_{ri} \tag{8.35}$$

where the shapes are written in homogeneous forms. The transformation $\mathbf{H}_{ri}$ can be linearly computed from four pairs of point correspondences in a general position between the two structures. However, as we do not know whether the selected features are rigid or nonrigid, we adopt an iterative RANSAC paradigm [6] to compute the transformation of rigid features.

During each iteration, we randomly draw four pairs of corresponding 3D features across the sequence and use these data sets to hypothesize the transformation matrix $\mathbf{H}_{ri}$ $(i = 2, \ldots, m)$. Then, we register all structures to the reference frame. Suppose $\bar{\mathbf{S}}_i'$ is the transformed structure of $\bar{\mathbf{S}}_i$. We define the registration error as the Euclidean distance between each pair of features.

$$\mathbf{e}_i = \mathrm{diag}\big((\bar{\mathbf{S}}_i' - \bar{\mathbf{S}}_1)^T(\bar{\mathbf{S}}_i' - \bar{\mathbf{S}}_1)\big) \tag{8.36}$$

$$\mathbf{E} = \frac{1}{m-1}\sum_{j=2}^{m}\mathbf{e}_i \tag{8.37}$$

where diag($\mathbf{X}$) stands for the main diagonal of the matrix $\mathbf{X}$; $\mathbf{e}_i \in \mathbb{R}^n$ stands for the error of each feature in frame i, and $\mathbf{E} \in \mathbb{R}^n$ stands for the mean registration error of each feature in all frames. Then the correct transformation can be chosen as the one with the most supporting features, i.e. small registration errors.

The RANSAC algorithm selects samples with uniform probability, which is computationally expensive especially for large data. Suppose the fraction of the rigid features is γ, then the trial number N can be determined from

$$N = \frac{\log(1-P)}{\log(1-\gamma^4)} \tag{8.38}$$

where P is the probability that the randomly selected 4 sets of samples are rigid features. For example, if we set $P = 0.99$, $\gamma = 0.4$. The trials will be $N = 178$. However, as we have obtained the mean reprojection error $\mathbf{e}_r$ in the initialization step of the last section, the efficiency of the searching process can be improved by incorporating this prior information. Intuitively, points with small reprojection errors are given higher drawing probability since they are more likely to be rigid, whereas features with large errors are given lower probability. Suppose the fraction of rigid features is increased to $\gamma' = 0.7$ by removing features with large reprojection errors. Then the trials will be reduced to $N' = 17$ under the same probability $P = 0.99$. Thus, more than 90% of the computational cost is saved.

After recovering the correct transformation matrix $\mathbf{H}_{ri}$, we can register all structures to the reference frame and compare their registration errors. Since the structure of the deformation part varies from frame to frame, the mean registration error of these features is much larger than that of the rigid ones. Therefore, the deformation is easily distinguished as shown in the tests. Previous methods tried to detect the deformation directly from 2D measurements [4, 5]. This is a complex problem as the constraints for segmentation in

2D are prone to be violated by noise. While in 3D space, we have a more geometrically meaningful information and the error of the deformation part is accumulated frame by frame. Thus, more accurate and robust segmentation results are expected.

8.4.2 Stratification to Euclidean Space

We have segmented the rigid features $\mathbf{S}_{ri}$ from the deformation ones $\mathbf{S}_{ni}$. For uncalibrated rigid part, there are many effective methods to recover the Euclidean shape and motions [7, 10, 11]. Thus, we obtain the Euclidean structure $\bar{\mathbf{S}}_{er}$ of the rigid part via one of these methods. From Corollary 8.1, we know that the affine solution can be stratified to the Euclidean space via affine transformation $\mathbf{H}_{ai}$, which can be computed as follows.

$$\tilde{\mathbf{S}}_{er} = \mathbf{H}_{ai}\tilde{\mathbf{S}}_{ri} \tag{8.39}$$

The stratification matrix in (8.39) is estimated only from the rigid part. Thus, the influence caused by large tracking errors of the nonrigid features may be relaxed, since feature tracking of deformable objects is difficult due to the absence of disambiguating geometric constraint. After $\mathbf{H}_{ai}$ is recovered, the deformation part as well as the whole structure can be stratified to the Euclidean space using

$$\tilde{\mathbf{S}}_{eni} = \mathbf{H}_{ai}\tilde{\mathbf{S}}_{ni}, \qquad \tilde{\mathbf{S}}_{ei} = \mathbf{H}_{ai}\tilde{\mathbf{S}}_{i} \tag{8.40}$$

where $\tilde{\mathbf{S}}_{eni}$ and $\tilde{\mathbf{S}}_{ei}$ stand for the Euclidean shape of the deformation part and the entire structure respectively. The solution of (8.40) is sub-optimal since the stratification matrix is computed using the rigid features only. Finally, a global optimization scheme is followed after stratification.

Suppose the Euclidean motion matrix and the structure after stratification are $\mathbf{A}_{ei}$ and $\tilde{\mathbf{S}}_{ei}$ respectively, these parameters will be further optimized by minimizing the image reprojection residuals.

$$J(\mathbf{A}_{ei}, \tilde{\mathbf{S}}_{ei}) = \min \left\| \mathbf{W} - \bar{\mathbf{W}} \right\|_F^2 \tag{8.41}$$

where $\bar{\mathbf{W}}$ denotes the reprojected tracking matrix. The minimization process is also termed as bundle adjustment in computer vision society that can be solved via Newton iteration or Levenberg-Marquardt iteration method [9].

8.4.3 Implementation Outline

The implementation details of the proposed method are summarized as follows.

1. For a given tracking matrix, perform rigid factorization and construct the initial values according to (8.31);
2. Compute the affine shape bases and structure according to the constrained power factorization algorithm;
3. Segment the rigid features and the deformation ones from the 3D affine structure via RANSAC algorithm;
4. Calculate the stratification matrix from (8.39) and stratify the structure from affine to Euclidean space;
5. Perform a global optimization of the recovered structure and motion by minimizing (8.41).

8.5 Evaluations on Synthetic Data

We generated a synthetic cube with three visible surfaces in space, with a dimension of $10 \times 10 \times 10$ with 9 evenly distributed points on each edge. There are three sets of moving points (17×3 points) on the adjacent surfaces of the cube that move at a constant speed as shown in Fig. 8.1. The object is composed of 90 rigid points and 51 deformation points. We generate 20 frames with different camera parameters by perspective projection. The image size is 500×500, while the distance of the camera to the object is set at about 120 to simulate affine imaging conditions.

8.5.1 Reconstruction Results and Evaluations

During the test, 1-pixel Gaussian noise was added to the images. We recover the 3D Euclidean structures of the 20 frames by the proposed stratification algorithm and automatically register all structures to the first frame via RANSAC. The result is shown in Fig. 8.1. We perform a comparison with the SVD-based method with rotation and basis constraints (SVD+RB) [18] as shown in Fig. 8.1. One may see from the results that the deformation structure is correctly recovered by both methods.

It should be noted that the recovered structures by the two methods are defined up to a 3D similarity transformation with the ground truth. For evaluation, we compute the similarity matrix and register the recovered structure with the ground truth. Then, we evaluate the reconstruction errors as point-to-point distances between the recovered structure and the ground truth. Figure 8.2 shows the mean and standard deviation of the errors associated with each frame at two different noise levels. As a comparison, we include results obtained by the SVD-based method only with a rotation constraint (SVD+R) [3] and the method with both rotation and basis constraints (SVD+RB) [18]. We can see from Fig. 8.2 that the proposed method outperforms the two SVD-based methods.

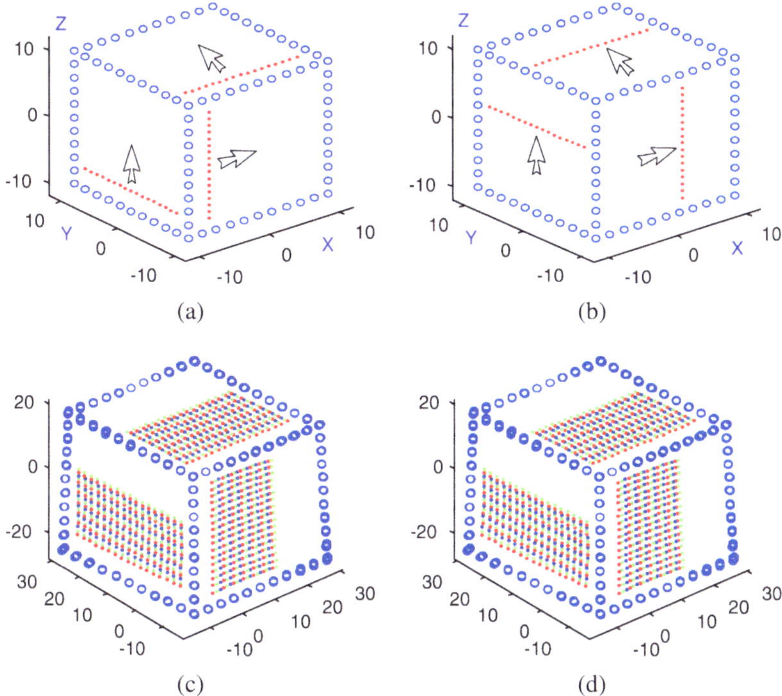

Fig. 8.1 Synthetic data and reconstruction results. (**a**) & (**b**) The synthetic cubes correspond to the first and last frames, where the rigid and moving points are denoted by circles and dots respectively; (**c**) The registered 3D structures of the 20 frames obtained by the proposed method; (**d**) The registered 3D structures by the method of SVD+RB

8.5.2 Convergence Property and Segmentation

We tested the convergence rate in three cases: First, we use all data without added noise. Second, we randomly delete 20% points from the tracking matrix. Third, we add 2-pixel Gaussian noise to the imaged features. At each iteration, we record the variation (8.33) of the reprojected tracking matrix and the relative reprojection error as defined by

$$\mathbf{E}_{rep} = \frac{\|\mathbf{W} - (\Omega_t \otimes \Phi_t)\bar{\mathbf{B}}_t\|_F^2}{\|\mathbf{W}\|_F^2} \times 100\ (\%) \tag{8.42}$$

The results are shown in Fig. 8.3. It is evident from these tests that the CPF algorithm converges quickly, even with some missing data and measurement errors. We compared the computation time of different methods on an Intel Pentium 4 3.6 GHz CPU programmed with Matlab 6.5. One iteration of the CPF algorithm takes 0.062 seconds, the SVD-based method [18] takes 0.017 seconds. Clearly, the CPF algorithm takes much more computa-

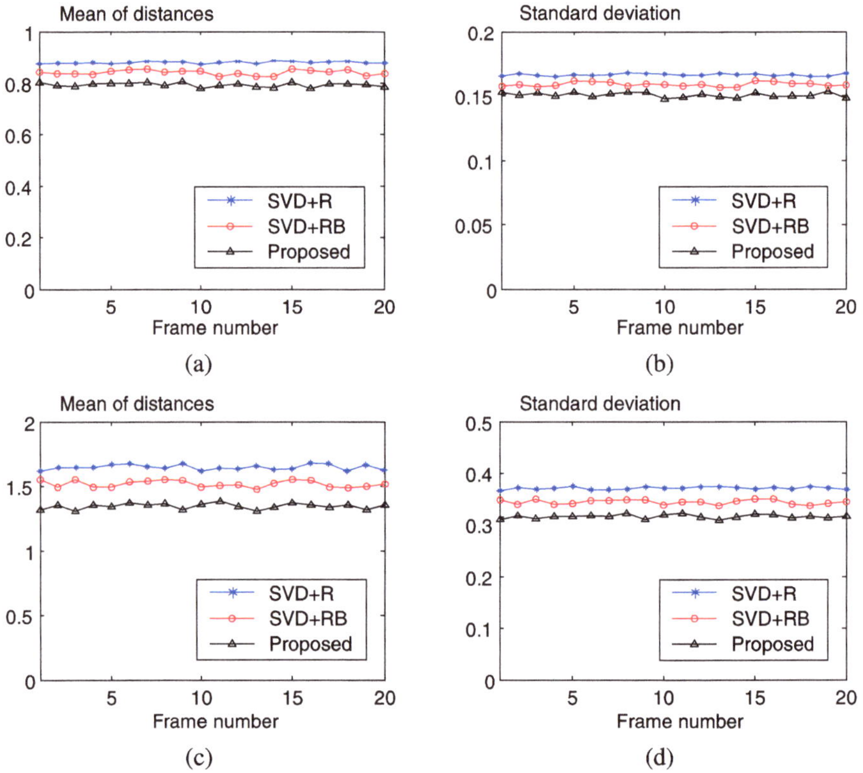

Fig. 8.2 Performance evaluation with ground truth. (**a**) & (**b**) The mean and standard deviation of the reconstruction errors with 1-pixel Gaussian noise; (**c**) & (**d**) The mean and standard deviation with 2-pixel Gaussian noise

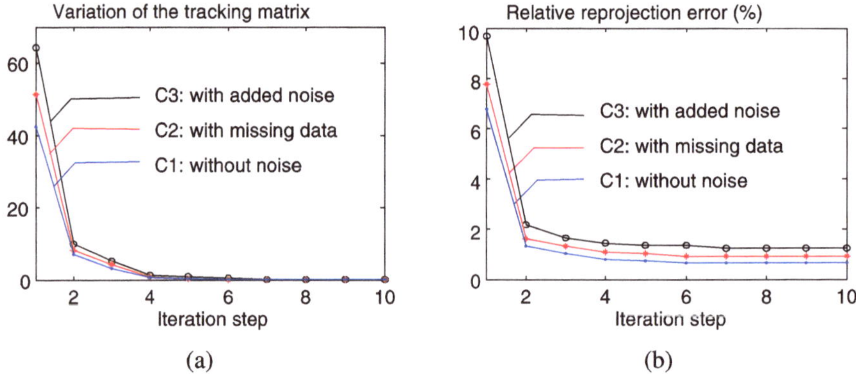

Fig. 8.3 Convergence property of the CPF algorithm in three conditions. (**a**) The variation of the reprojected tracking matrix at each iteration; (**b**) The relative reprojection error at each iteration

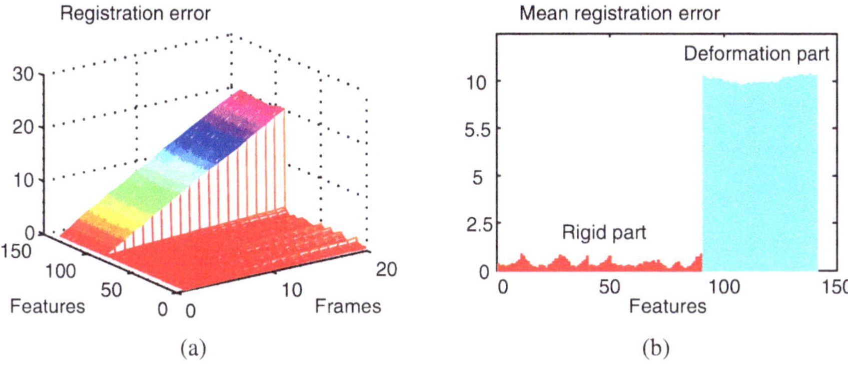

Fig. 8.4 Segmentation result. (**a**) The registration error of the features in each frame; (**b**) The mean registration error of every feature across the sequence

Table 8.1 Misclassification error with respect to different noise levels and deformation/rigid feature ratios out of 100 trials

Noise level		0	0.5	1	1.5	2
	51/90	0.00	0.00	0.05	0.37	0.59
Ratio	51/60	0.00	0.06	0.15	0.42	0.83
	51/30	0.09	0.21	0.54	1.05	1.77

tion time than the SVD-based method. However, the algorithm is still very fast for most applications.

From the CPF algorithm, we recover the affine structure and shape bases, and register all structures to the first view automatically via RANSAC algorithm. The registration error (8.36) of the features in each frame and the mean error (8.37) of every features across the sequence are shown in Fig. 8.4, where the first 90 features belong to the rigid part. We observe that it is easy to detect and segment the deformation part by virtue of the mean registration error in 3D space.

The detection strategy may be affected by the ratio of nonrigid features to the rigid ones, noise level, threshold value, deformation amplitude, etc. We studied the misclassification error (the number of misclassified features) with respect to noise level and the ratio of nonrigid features. The results are tabulated in Table 8.1, where we vary the number of rigid features from 90 to 30, while the number of nonrigid features are fixed to 51; the noise level is varied from 0 to 2 pixels. The values in Table 8.1 are evaluated from 100 independent tests. In real applications, the threshold is determined experimentally based on the distribution of the mean registration error (8.37). We usually avoid the misclassification of deformable features into rigid ones by reducing the threshold such that the stratification matrices may be recovered more accurately.

8

8.6 Evaluations on Real Sequences

We tested the proposed methods on several real image sequences and report two results in this section. The two sequences in the test were captured by a Canon Powershot G3 digital camera.

8.6.1 Test on Grid Sequence

The sequence is similar to that used in Chap. 7. Two frames are shown in Fig. 8.5. The background of the sequence consists of two orthogonal sheets with square grids which are

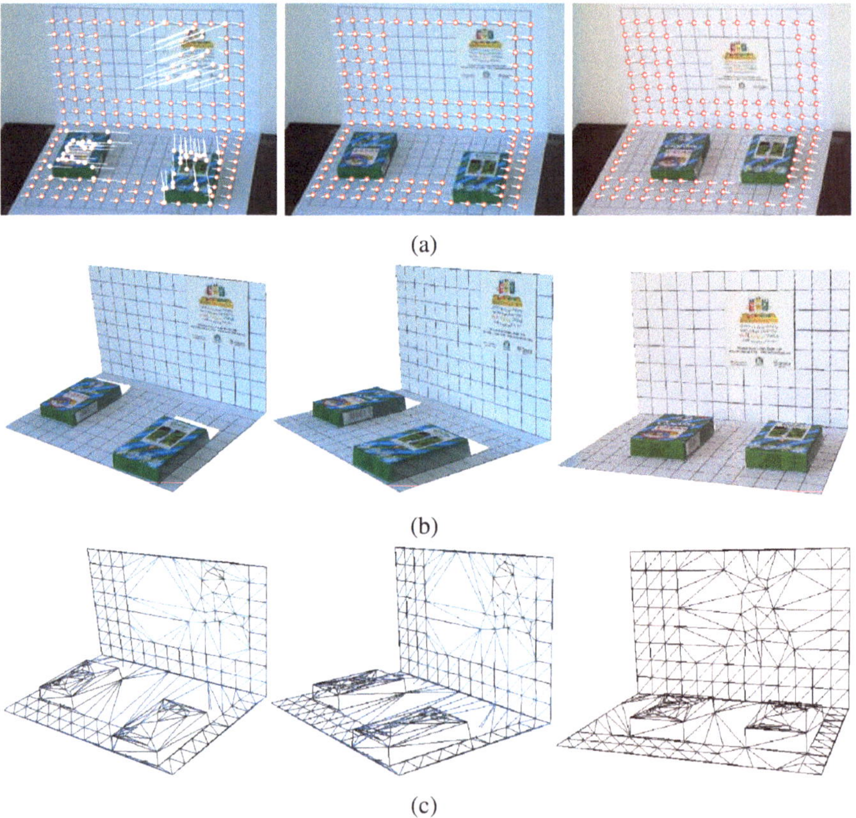

Fig. 8.5 Reconstruction results of grid sequence. (**a**) Two frames from the sequence overlaid with 206 tracked features and 140 automatically segmented rigid features with relative disparities shown in *white lines*; (**b**) The reconstructed VRML models of the two frames shown in different viewpoints with texture mapping; (**c**) The corresponding wireframes of the VRML models

Table 8.2 Performance comparison and evaluation of the proposed method with respect to the SVD-based methods

Sequence	Method	$\mathbf{E}_{angb}$	$\mathbf{E}_{ratio}$	$\mathbf{E}_{angd}$	$\mathbf{E}_{rep}$
Grid	Proposed	0.025	0.108	0.012	4.293
	SVD+RB	0.041	0.147	0.016	5.662
	SVD+R	0.049	0.153	0.024	5.817

used as ground truth for evaluation. A total of 206 tracked features are established interactively across the sequence, where 140 features belong to the static background and 66 features belong to the three moving objects. All static and moving features are automatically separated by the proposed RANSAC scheme as shown in Fig. 8.5. We recovered the metric structure of the scenario by the proposed method. The reconstructed VRML models and corresponding triangulated wireframes of two frames at different viewpoints are shown in Fig. 8.5. The dynamic structure of scenario is correctly recovered.

For more performance evaluation and comparison, we compared the relative reprojection errors $\mathbf{E}_{rep}$ by the proposed method and the two SVD-based methods. Then, we computed the angle between the two orthogonal background sheets, the length ratio of the two diagonals of each square and the angle formed by the two diagonals. The mean errors of these three values are denoted by $\mathbf{E}_{angb}$, $\mathbf{E}_{ratio}$ and $\mathbf{E}_{angd}$ respectively. The comparative results obtained by different methods are listed in Table 8.2. We can see that the proposed method performs better than the SVD-based methods as expected.

8.6.2 Test on Toy Sequence

There are 25 frames in the sequence and the image resolution is 1024×768. The scene is composed of three rigid objects, where the clock tower is fixed and the two clay babies move slowly during shooting. We established the initial correspondences by the method in [20] and interactively deleted some outliers. The feature tracking is hard for this sequence due to the existence of smooth texture on the surface of the clay babies. Figure 8.6 shows the detected and matched features of two frames. In this test, instead of using the features being tracked across the entire sequence, we utilize only those features that are tracked across more than 20 frames. Thus, there is about 12% missing data in the tracking matrix. We perform the CPF algorithm on the tracking matrix and recover the affine structure of the scene. Then, we segment the static and moving features and upgrade the solution to the Euclidean space. Figure 8.6 shows the reconstructed 3D structures and wireframes of the frames from different viewpoints. One may notice from the result that some 3D points are not accurately recovered due to the tracking errors. However, the structures of the scene are largely reasonable. The relative reprojection error by the proposed method is 7.164. We do not have the results of the SVD-based algorithm for this sequence due to incomplete tracking matrix.

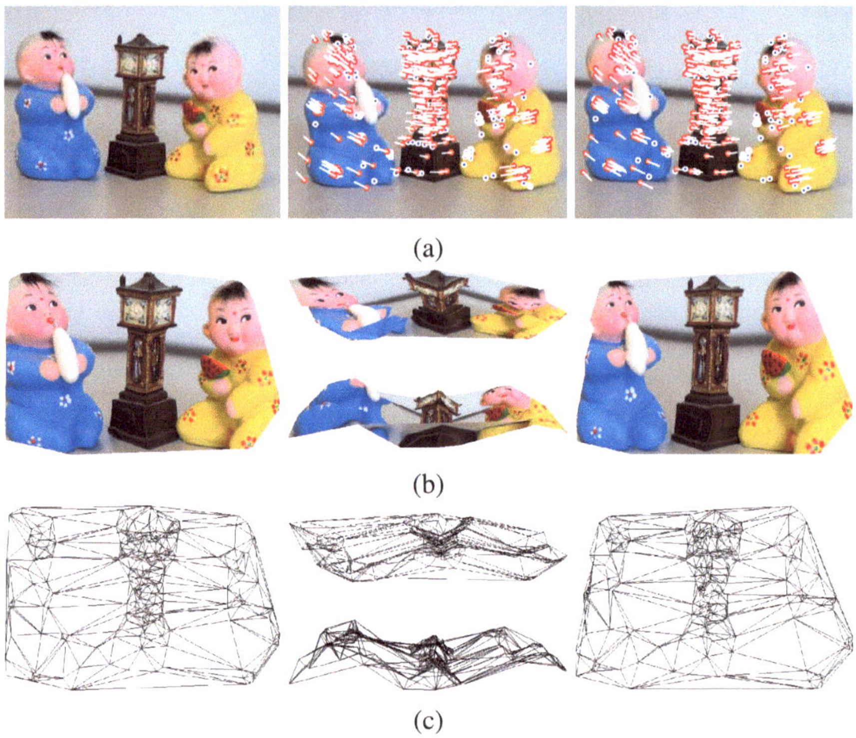

Fig. 8.6 Reconstruction results of toy sequence with 12% missing data. (**a**) Three frames from the sequence with 427 detected features overlaid on the second frame and 439 features on the third frame. Among them 352 points are tracked correctly as shown with relative disparities in *white lines*, the other not matched points are shown in *black dots with white circles*; (**b**) The reconstructed VRML models shown in different viewpoints with texture mapping; (**c**) The corresponding wireframes of the reconstructed models

8.7 Closure Remarks

8.7.1 Conclusion

In this chapter, we first proposed a deformation weight constraint to ensure the invariant relationship between the recovered shape bases and structures. Then we presented the CPF algorithm to recover the deformation structure in affine space. Based on the 3D affine structures, we proposed a RANSAC based strategy to detect and segment the rigid features from the deformation ones, and stratified the solution from affine to the Euclidean space by virtue of the rigid features. This is an alternative method for nonrigid factorization which avoids the difficulty in recovering the upgrading matrix. Experiments on synthetic and real sequences show improvements over SVD-based methods.

8.7.2 Review Questions

1. *Deformation weight constraint.* Give the deformation weight constraint and deduce the invariant relationship in (8.1) under Euclidean and affine transformations. Will the constraint affect previous SVD-based nonrigid factorization?
2. *CPF algorithm.* Provide computational details of the constrained power factorization. Compare CPF to the RCPF algorithm discussed in the last chapter? How to deal with missing data in the tracking matrix?
3. *Deformation detection.* Explain the strategy involved in the proposed segmentation method. How can you improve the efficiency of the RANSAC process by incorporating the mean reprojection error?
4. *Euclidean Stratification.* Elucidate the computation of affine transformation matrix in order to stratify the solution from affine to Euclidean? How to compute the reprojected tracking matrix in the optimization process?

References

1. Brand, M.: Morphable 3D models from video. In: Proc. of IEEE Conference on Computer Vision and Pattern Recognition, vol. 2, pp. 456–463 (2001)
2. Brand, M.: A direct method for 3D factorization of nonrigid motion observed in 2D. In: Proc. of IEEE Conference on Computer Vision and Pattern Recognition, vol. 2, pp. 122–128 (2005)
3. Bregler, C., Hertzmann, A., Biermann, H.: Recovering non-rigid 3D shape from image streams. In: Proc. of IEEE Conference on Computer Vision and Pattern Recognition, vol. 2, pp. 690–696 (2000)
4. Del Bue, A., Lladó, X., de Agapito, L.: Non-rigid face modelling using shape priors. In: Proc. of Second International Workshop on Analysis and Modelling of Faces and Gestures, pp. 97–108 (2005)
5. Del Bue, A., Lladó, X., de Agapito, L.: Non-rigid metric shape and motion recovery from uncalibrated images using priors. In: Proc. of IEEE Conference on Computer Vision and Pattern Recognition, vol. 1, pp. 1191–1198 (2006)
6. Fischler, M.A., Bolles, R.C.: Random sample consensus: A paradigm for model fitting with applications to image analysis and automated cartography. Commun. ACM **24**(6), 381–395 (1981)
7. Han, M., Kanade, T.: Creating 3D models with uncalibrated cameras. In: Proc. of IEEE Computer Society Workshop on the Application of Computer Vision (2000)
8. Hartley, R., Schaffalizky, F.: Powerfactorization: 3D reconstruction with missing or uncertain data. In: Proc. of Australia-Japan Advanced Workshop on Computer Vision (2003)
9. Hartley, R.I., Zisserman, A.: Multiple View Geometry in Computer Vision, 2nd edn. Cambridge University Press, Cambridge (2004). ISBN: 0521540518
10. Poelman, C., Kanade, T.: A paraperspective factorization method for shape and motion recovery. IEEE Trans. Pattern Anal. Mach. Intell. **19**(3), 206–218 (1997)
11. Quan, L.: Self-calibration of an affine camera from multiple views. Int. J. Comput. Vis. **19**(1), 93–105 (1996)
12. Torresani, L., Hertzmann, A., Bregler, C.: Learning non-rigid 3D shape from 2D motion. In: Proc. of Advances in Neural Information Processing Systems (2004)
13. Torresani, L., Hertzmann, A., Bregler, C.: Nonrigid structure-from-motion: Estimating shape and motion with hierarchical priors. IEEE Trans. Pattern Anal. Mach. Intell. **30**(5), 878–892 (2008)

14. Torresani, L., Yang, D.B., Alexander, E.J., Bregler, C.: Tracking and modeling non-rigid objects with rank constraints. In: Proc. of IEEE Conference on Computer Vision and Pattern Recognition, vol. 1, pp. 493–500 (2001)
15. Wang, G., Tian, Y., Sun, G.: Modelling nonrigid object from video sequence under perspective projection. In: Lecture Notes in Computer Science, vol. 3784, pp. 64–71 (2005)
16. Wang, G., Tsui, H.T., Wu, J.: Rotation constrained power factorization for structure from motion of nonrigid objects. Pattern Recogn. Lett. **29**(1), 72–80 (2008)
17. Wang, G., Wu, J.: Stratification approach for 3-D Euclidean reconstruction of nonrigid objects from uncalibrated image sequences. IEEE Trans. Syst. Man Cybern., Part B **38**(1), 90–101 (2008)
18. Xiao, J., Chai, J., Kanade, T.: A closed-form solution to non-rigid shape and motion recovery. Int. J. Comput. Vis. **67**(2), 233–246 (2006)
19. Xiao, J., Kanade, T.: Uncalibrated perspective reconstruction of deformable structures. In: Proc. of the International Conference on Computer Vision, vol. 2, pp. 1075–1082 (2005)
20. Yao, J., Cham, W.: Feature matching and scene reconstruction from multiple widely separated views. Tech. Rep., Chinese University of Hong Kong (2005)

Quasi-Perspective Factorization 9

Abstract Previous studies on structure and motion factorization are either based on simplified affine assumption or general perspective projection. The affine approximation is widely adopted due to its simplicity, whereas the extension to perspective model suffers from difficulties in projective depth recovery. To fill the gap between simplicity of affine and accuracy of perspective model, we propose a quasi-perspective factorization algorithm for structure and motion recovery of both rigid and nonrigid objects. Firstly, we establish a framework of rigid and nonrigid factorization under quasi-perspective assumption. Secondly, we propose an extended Cholesky decomposition to recover the rotation part of the Euclidean upgrading matrix. Finally, we prove that the last column of the upgrading matrix corresponds to a global scale and translation of the camera thus may be set freely. The proposed algorithm is validated and evaluated extensively on synthetic and real image sequences.

Each problem that I solved became a rule which served afterwards to solve other problems.

René Descartes (1596–1650)

9.1 Introduction

The factorization algorithm is a powerful and efficient method for structure and motion recovery. Since Tomasi and Kanade [17] firstly introduced the algorithm in the early 90's, numerous extensions and generalizations have been proposed. Most early studies on the problem assume rigid object and affine camera model [13, 14]. The main difficulty of its extension to perspective projection lies in the recovery of depth scales. One method is to estimate the depths in a pair-wise fashion via epipolar constraint [16, 19], which may be unstable due to possible error accumulation. Another method is based on nonlinear iteration by minimizing reprojections [5, 8, 10]. These methods rely on the accuracy of initial affine solution. Oliensis and Hartley [11] recently proved that no iteration converges

G. Wang, Q.M.J. Wu, *Guide to Three Dimensional Structure and Motion Factorization*, Advances in Pattern Recognition,
DOI 10.1007/978-0-85729-046-5_9,

9

sensibly. Wang and Wu [24] proposed a hybrid method to initialize the depth scales via a projective reconstruction.

In recent years, many extensions stemming from the factorization algorithm were proposed to relax the rigidity constraint to multiple moving objects [3, 9] and articulated objects [29, 30]. Bregler *et al.* [2] firstly established the framework of nonrigid factorization using shape bases. The method was extensively investigated and developed under affine assumption [1, 4, 18, 27]. It was extended to perspective projection in [21, 28]. Rabaud and Belongie [15] relaxed the Bregler's assumption and proposed to solve the problem by a manifold-learning framework. Wang *et al.* [22] introduced a rotation constrained power factorization algorithm. Hartley and Vidal [7] proposed a closed form solution to the nonrigid shape and motion with known camera constraints.

The affine camera model is widely adopted in factorization due to its simplicity. However, the accuracy of this approximation is not satisfactory in many applications. Perspective projection based algorithm is computationally intensive and its convergence is not guaranteed. In this chapter, we will apply the quasi-perspective projection model [23] for both rigid and nonrigid factorization framework. This is a trade-off between the simplicity of affine and accuracy of full perspective projection. It is proved to be more accurate than affine approximation since the projective depths in quasi-perspective projection are implicitly embedded in the motion and shape matrices. While the difficult problem of depth recovery in perspective factorization is avoided [26].

The remaining part of this chapter is organized as follows. The factorization algorithm is briefly reviewed in Sect. 9.2. The proposed quasi-perspective factorization algorithm for rigid objects is detailed in Sect. 9.3. The nonrigid factorization under quasi-perspective projection is presented in Sect. 9.4. Some experimental evaluations on synthetic and real image sequences are reported in Sects. 9.5 and 9.6 respectively.

9.2 Background on Factorization

We already introduced the rigid and nonrigid factorization under affine and perspective projection models in the previous chapters. For convenience of discussion, we present a brief review on the expressions of different factorization algorithms.

Under perspective projection, a 3D point $\mathbf{X}_j$ is imaged at $\mathbf{x}_{ij}$ in the ith frame according to equation

$$\lambda_{ij}\mathbf{x}_{ij} = \mathbf{P}_i\mathbf{X}_j = \mathbf{K}_i[\mathbf{R}_i, \mathbf{T}_i]\mathbf{X}_j \tag{9.1}$$

If we adopt affine projection model and register all image points to the centroid. Then, the projection process (9.1) is simplified to the form

$$\bar{\mathbf{x}}_{ij} = \mathbf{A}_i\bar{\mathbf{X}}_j \tag{9.2}$$

Given n tracked features of an object across a sequence of m frames. The structure and motion factorization under affine assumption (9.2) can be expressed as

$$\underbrace{\begin{bmatrix} \bar{\mathbf{x}}_{11} & \cdots & \bar{\mathbf{x}}_{1n} \\ \vdots & \ddots & \vdots \\ \bar{\mathbf{x}}_{m1} & \cdots & \bar{\mathbf{x}}_{mn} \end{bmatrix}}_{\mathbf{W}_{2m\times n}} = \underbrace{\begin{bmatrix} \mathbf{A}_1 \\ \vdots \\ \mathbf{A}_m \end{bmatrix}}_{\mathbf{M}_{2m\times 3}} \underbrace{[\bar{\mathbf{X}}_1, \ldots, \bar{\mathbf{X}}_n]}_{\bar{\mathbf{S}}_{3\times n}} \tag{9.3}$$

The factorization is usually performed by SVD decomposition of the tracking matrix $\mathbf{W}$ with rank-3 constraint. When the perspective projection model (9.1) is adopted, the factorization can be modeled as

$$\underbrace{\begin{bmatrix} \lambda_{11}\mathbf{x}_{11} & \cdots & \lambda_{1n}\mathbf{x}_{1n} \\ \vdots & \ddots & \vdots \\ \lambda_{m1}\mathbf{x}_{m1} & \cdots & \lambda_{mn}\mathbf{x}_{mn} \end{bmatrix}}_{\dot{\mathbf{W}}_{3m\times n}} = \underbrace{\begin{bmatrix} \mathbf{P}_1 \\ \vdots \\ \mathbf{P}_m \end{bmatrix}}_{\mathbf{M}_{3m\times 4}} \underbrace{\begin{bmatrix} \bar{\mathbf{X}}_1, & \ldots, & \bar{\mathbf{X}}_n \\ 1, & \ldots, & 1 \end{bmatrix}}_{\mathbf{S}_{4\times n}} \tag{9.4}$$

The rank of the projective depth-scaled tracking matrix $\dot{\mathbf{W}}$ is at most 4 if a consistent set of scalars are present.

When an object is nonrigid, we follow Bregler's assumption [2] to model the nonrigid structure by a linearly combination of some shape bases.

$$\bar{\mathbf{S}}_i = \sum_{l=1}^{k} \omega_{il}\mathbf{B}_l \tag{9.5}$$

With this assumption, the nonrigid factorization under affine camera model is expressed as

$$\underbrace{\begin{bmatrix} \bar{\mathbf{x}}_{11} & \cdots & \bar{\mathbf{x}}_{1n} \\ \vdots & \ddots & \vdots \\ \bar{\mathbf{x}}_{m1} & \cdots & \bar{\mathbf{x}}_{mn} \end{bmatrix}}_{\mathbf{W}_{2m\times n}} = \underbrace{\begin{bmatrix} \omega_{11}\mathbf{A}_1 & \cdots & \omega_{1k}\mathbf{A}_1 \\ \vdots & \ddots & \vdots \\ \omega_{m1}\mathbf{A}_m & \cdots & \omega_{mk}\mathbf{A}_m \end{bmatrix}}_{\mathbf{M}_{2m\times 3k}} \underbrace{\begin{bmatrix} \mathbf{B}_1 \\ \vdots \\ \mathbf{B}_k \end{bmatrix}}_{\bar{\mathbf{B}}_{3k\times n}} \tag{9.6}$$

and the rank of the nonrigid tracking matrix is at most $3k$. Similarly, the factorization under perspective projection can be formulated as follows [28].

$$\underbrace{\begin{bmatrix} \lambda_{11}\mathbf{x}_{11} & \cdots & \lambda_{1n}\mathbf{x}_{1n} \\ \vdots & \ddots & \vdots \\ \lambda_{m1}\mathbf{x}_{m1} & \cdots & \lambda_{mn}\mathbf{x}_{mn} \end{bmatrix}}_{\dot{\mathbf{W}}_{3m\times n}} = \underbrace{\begin{bmatrix} \omega_{11}\mathbf{P}_1^{(1:3)} & \cdots & \omega_{1k}\mathbf{P}_1^{(1:3)} & \mathbf{P}_1^{(4)} \\ \vdots & \ddots & \vdots & \vdots \\ \omega_{m1}\mathbf{P}_m^{(1:3)} & \cdots & \omega_{mk}\mathbf{P}_m^{(1:3)} & \mathbf{P}_m^{(4)} \end{bmatrix}}_{\mathbf{M}_{3m\times(3k+1)}} \underbrace{\begin{bmatrix} \mathbf{B}_1 \\ \vdots \\ \mathbf{B}_k \\ \mathbf{1}^T \end{bmatrix}}_{\mathbf{B}_{(3k+1)\times n}} \tag{9.7}$$

The rank of the correctly scaled tracking matrix in (9.7) is at most $3k+1$. Just as its rigid counterpart, the most difficult issue for perspective factorization is to determine the projective depths that are consistent with (9.1).

9.3 Quasi-Perspective Rigid Factorization

Under the assumption that the camera is far away from the object with small lateral rotations, we proposed a quasi-perspective projection model to simplify the imaging process as follows.

$$\mathbf{x}_{ij} = \mathbf{P}_{qi}\mathbf{X}_{qj} = (\mu_i \mathbf{P}_i)(\ell_j \mathbf{X}_j) \tag{9.8}$$

In quasi-perspective projection (9.8), the projective depths are implicitly embedded in the scalars of the homogeneous structure $\mathbf{X}_{qj}$ and the projection matrix $\mathbf{P}_{qi}$. Thus, the difficult problem of estimating the unknown depths is avoided. The model is more general than affine projection model (9.2), where all projective depths are simply assumed to be equal. Under the quasi-perspective assumption, the factorization equation of a tracking matrix is expressed as

$$\begin{bmatrix} \mathbf{x}_{11} & \cdots & \mathbf{x}_{1n} \\ \vdots & \ddots & \vdots \\ \mathbf{x}_{m1} & \cdots & \mathbf{x}_{mn} \end{bmatrix} = \begin{bmatrix} \mu_1 \mathbf{P}_1 \\ \vdots \\ \mu_m \mathbf{P}_m \end{bmatrix} [\ell_1 \mathbf{X}_1, \ldots, \ell_n \mathbf{X}_n] \tag{9.9}$$

which can be written concisely as

$$\tilde{\mathbf{W}}_{3m\times n} = \mathbf{M}_{3m\times 4}\mathbf{S}_{4\times n} \tag{9.10}$$

The form is similar to perspective factorization (9.4). However, the projective depths in (9.9) are embedded in the motion and shape matrices, hence there is no need to estimate them explicitly. By performing SVD on the tracking matrix and imposing the rank-4 constraint, $\tilde{\mathbf{W}}$ may be factorized as $\hat{\mathbf{M}}_{3m\times 4}\hat{\mathbf{S}}_{4\times n}$. However, like all other factorization algorithms, the decomposition is not unique since it is defined up to a nonsingular linear transformation $\mathbf{H}_{4\times 4}$ as $\mathbf{M} = \hat{\mathbf{M}}\mathbf{H}$ and $\mathbf{S} = \mathbf{H}^{-1}\hat{\mathbf{S}}$. Due to the special form of (9.9), the upgrading matrix has some special properties compared to that under affine and perspective projection. We will present the computational details in the following section.

9.3.1 Euclidean Upgrading Matrix

We adopt the metric constraint to compute an upgrading matrix $\mathbf{H}_{4\times 4}$. Let us decompose the matrix into two parts as

$$\mathbf{H} = [\mathbf{H}_l | \mathbf{H}_r] \tag{9.11}$$

where $\mathbf{H}_l$ denotes the first three columns, and $\mathbf{H}_r$ denotes the fourth column. Suppose $\hat{\mathbf{M}}_i$ is the ith triple rows of $\hat{\mathbf{M}}$, then we have

$$\hat{\mathbf{M}}_i\mathbf{H} = [\hat{\mathbf{M}}_i\mathbf{H}_l|\hat{\mathbf{M}}_i\mathbf{H}_r] \tag{9.12}$$

where

$$\hat{\mathbf{M}}_i\mathbf{H}_l = \mu_i\mathbf{P}_i^{(1:3)} = \mu_i\mathbf{K}_i\mathbf{R}_i \tag{9.13}$$

$$\hat{\mathbf{M}}_i\mathbf{H}_r = \mu_i\mathbf{P}_i^{(4)} = \mu_i\mathbf{K}_i\mathbf{T}_i \tag{9.14}$$

where $\mathbf{P}_i^{(1:3)}$ and $\mathbf{P}_i^{(4)}$ denote the first three columns and the fourth column of $\mathbf{P}_i$. In the following, we will show how to compute $\mathbf{H}_l$ and $\mathbf{H}_r$.

9.3.1.1 Recovering $\mathbf{H}_l$

Let us denote $\mathbf{Q} = \mathbf{H}_l\mathbf{H}_l^T$, which is a 4×4 symmetric matrix. As in previous factorization studies [5, 14], we adopt a simplified camera model with only one parameter as $\mathbf{K}_i = \mathrm{diag}(f_i, f_i, 1)$. Then, from

$$\begin{aligned}\mathbf{C}_i &= \hat{\mathbf{M}}_i\mathbf{Q}\hat{\mathbf{M}}_i^T = \left(\hat{\mathbf{M}}_i\mathbf{H}_l\right)\left(\hat{\mathbf{M}}_i\mathbf{H}_l\right)^T = (\mu_i\mathbf{K}_i\mathbf{R}_i)(\mu_i\mathbf{K}_i\mathbf{R}_i)^T \\ &= \mu_i^2\mathbf{K}_i\mathbf{K}_i^T = \mu_i^2\begin{bmatrix} f_i^2 & & \\ & f_i^2 & \\ & & 1\end{bmatrix}\end{aligned} \tag{9.15}$$

we obtain the following constraints.

$$\begin{cases}\mathbf{C}_i(1,2) = \mathbf{C}_i(2,1) = 0 \\ \mathbf{C}_i(1,3) = \mathbf{C}_i(3,1) - 0 \\ \mathbf{C}_i(2,3) = \mathbf{C}_i(3,2) = 0 \\ \mathbf{C}_i(1,1) - \mathbf{C}_i(2,2) = 0\end{cases} \tag{9.16}$$

Since the factorization (9.9) can be defined up to a global scalar as $\tilde{\mathbf{W}} = \mathbf{MS} = (\varepsilon\mathbf{M})(\mathbf{S}/\varepsilon)$, we set $\mu_1 = 1$ to avoid the trivial solution of $\mathbf{Q} = \mathbf{0}$. Thus we have $4m + 1$ linear constraints in total on the 10 unknowns of $\mathbf{Q}$, which can be solved via least squares. Ideally, $\mathbf{Q}$ is a positive semidefinite symmetric matrix, the matrix $\mathbf{H}_l$ can be recovered from $\mathbf{Q}$ via extended Cholesky decomposition as follows.

Definition 9.1 (Vertical extended upper triangular matrix) Suppose $\mathbf{U}$ is a $n \times k$ $(n > k)$ matrix. We call $\mathbf{U}$ a vertical extended upper triangular matrix if it is of the form

$$\mathbf{U}_{ij} = \begin{cases} u_{ij} & \text{if } i \leq j + (n-k) \\ 0 & \text{if } i > j + (n-k)\end{cases} \tag{9.17}$$

9

where $\mathbf{U}_{ij}$ denotes the (i,j)th element of $\mathbf{U}$, and u_{ij} is a scalar. For example, a $n \times (n-1)$ vertical extended upper triangular matrix can be written explicitly as

$$\mathbf{U} = \begin{bmatrix} u_{11} & u_{12} & \cdots & u_{1(n-1)} \\ u_{21} & u_{22} & \cdots & u_{2(n-1)} \\ & u_{32} & \cdots & u_{3(n-1)} \\ & & \ddots & \vdots \\ & & & u_{n(n-1)} \end{bmatrix} \tag{9.18}$$

Proposition 9.1 (Extended Cholesky Decomposition) *Suppose* $\mathbf{Q}_n$ *is a* $n \times n$ *positive semidefinite symmetric matrix of rank* k $(k < n)$. *Then it can be decomposed as* $\mathbf{Q}_n = \mathbf{H}_k\mathbf{H}_k^T$, *where* $\mathbf{H}_k$ *is a* $n \times k$ *matrix of rank* k. *Furthermore, the decomposition can be written as* $\mathbf{Q}_n = \Lambda_k\Lambda_k^T$ *with* Λ_k, *a* $n \times k$ *vertical extended upper triangular matrix. The degree-of-freedom of the matrix* $\mathbf{Q}_n$ *is* $nk - \frac{1}{2}k(k-1)$, *which is the number of unknowns in* Λ_k.

The proof is given in Appendix B. The proposition can be taken as an extension of Cholesky Decomposition to the case of positive semidefinite symmetric matrix, while Cholesky Decomposition can only deal with positive definite symmetric matrix. From the proposition, we obtain the following result.

Result 9.1 *The matrix* $\mathbf{Q}$ *recovered from* (9.16) *is a* 4×4 *positive semidefinite symmetric matrix of rank* 3. *It can be decomposed as* $\mathbf{Q} = \mathbf{H}_l\mathbf{H}_l^T$, *where* $\mathbf{H}_l$ *is a* 4×3 *rank* 3 *matrix. The decomposition can be further written as* $\mathbf{Q} = \Lambda_3\Lambda_3^T$ *with* Λ_3 *a* 4×3 *vertical extended upper triangular matrix.*

The computation of $\mathbf{H}_l$ is very simple. Suppose the SVD decomposition of $\mathbf{Q}$ is $\mathbf{U}_4\Sigma_4\mathbf{U}_4^T$, where $\mathbf{U}_4$ is a 4×4 orthogonal matrix, $\Sigma_4 = \mathrm{diag}(\sigma_1, \sigma_2, \sigma_3, 0)$ is a diagonal matrix with σ_i the singular value of $\mathbf{Q}$. Thus we have

$$\mathbf{H}_l = \mathbf{U}^{(1:3)} \begin{bmatrix} \sqrt{\sigma_1} & & \\ & \sqrt{\sigma_2} & \\ & & \sqrt{\sigma_3} \end{bmatrix} \tag{9.19}$$

Then the vertical extended upper triangular matrix Λ_3 can be constructed from $\mathbf{H}_l$ as in (B.26). From the number of unknowns in Λ_3, we know that $\mathbf{Q}$ is only defined up to 9 degrees of freedom.

In Result 9.1, we claim that the symmetric matrix $\mathbf{Q}$ can be decomposed into $\Lambda_3\Lambda_3^T$. In practice, we can simply decompose the matrix into $\mathbf{H}_l\mathbf{H}_l^T$ as shown in (9.19), it is unnecessary to recover Λ_3 since the upgrading matrix (9.11) is not unique. However, when the data is corrupted by noise, the recovered matrix $\mathbf{Q}$ may be negative definite and the decomposition of (9.19) is impossible. In such cases, we suggest the following alternative estimation method.

Let us denote

$$\Lambda_3 = \begin{bmatrix} h1 & h2 & h3 \\ h4 & h5 & h6 \\ & h7 & h8 \\ & & h9 \end{bmatrix} \tag{9.20}$$

and substitute the matrix $\mathbf{Q}$ in (9.15) with $\Lambda_3\Lambda_3^T$. Then a best estimation of Λ_3 in (9.20) can be obtained via minimizing the following cost function

$$J_1 = \min_{(\Lambda_3)} \frac{1}{2} \sum_{i=1}^{m} \left(\mathbf{C}_i^2(1,2) + \mathbf{C}_i^2(1,3) + \mathbf{C}_i^2(2,3) + (\mathbf{C}_i(1,1) - \mathbf{C}_i(2,2))^2 \right) \tag{9.21}$$

The minimization scheme can be solved using any nonlinear optimization techniques, such as gradient descent or Levenberg-Marquardt (LM) algorithm. By introducing the vertical extended upper triangular matrix (9.20), we can reduce three unknowns in matrix $\mathbf{Q}$.

9.3.1.2 Recovering $\mathbf{H}_r$

In this section we recover the right part $\mathbf{H}_r$ of the upgrading matrix (9.11). From quasi-perspective equation (9.8), we have

$$\mathbf{x}_{ij} = (\mu_i \mathbf{P}_i^{(1:3)})(\ell_j \bar{\mathbf{X}}_j) + (\mu_i \mathbf{P}_i^{(4)})\ell_j \tag{9.22}$$

For all features in the ith frame, we take a summation of their coordinates to get

$$\sum_{j=1}^{n} \mathbf{x}_{ij} = \mu_i \mathbf{P}_i^{(1:3)} \sum_{j=1}^{n} (\ell_j \bar{\mathbf{X}}_j) + \mu_i \mathbf{P}_i^{(4)} \sum_{j=1}^{n} \ell_j \tag{9.23}$$

where $\mu_i \mathbf{P}_i^{(1:3)}$ can be recovered from $\hat{\mathbf{M}}_i \mathbf{H}_l$, $\mu_i \mathbf{P}_i^{(4)} = \hat{\mathbf{M}}_i \mathbf{H}_r$. Since the world coordinate system can be chosen freely, we may set the origin of world system at the gravity center of the scaled space points as

$$\sum_{j=1}^{n} (\ell_j \bar{\mathbf{X}}_j) = 0 \tag{9.24}$$

On other hand, we may simply set

$$\sum_{j=1}^{n} \ell_j = 1 \tag{9.25}$$

since the reconstruction is defined up to a global scalar. Thus (9.23) is simplified to

$$\hat{\mathbf{M}}_i \mathbf{H}_r = \sum_{j=1}^{n} \mathbf{x}_{ij} = \begin{bmatrix} \sum_j u_{ij} \\ \sum_j v_{ij} \\ n \end{bmatrix} \tag{9.26}$$

which provides 3 linear constraints on the four unknowns of $\mathbf{H}_r$. Therefore, we obtain $3m$ equations from the sequence and $\mathbf{H}_r$ can be recovered via linear least squares.

From the above analysis, we note that the solution of $\mathbf{H}_r$ is not unique as it is dependant on selection of the world origin $\sum_{j=1}^{n}(\ell_j\bar{\mathbf{X}}_j)$ and the global scalar $\sum_{j=1}^{n}\ell_j$. Actually, $\mathbf{H}_r$ may be set freely as shown in the following proposition.

Proposition 9.2 *Suppose* $\mathbf{H}_l$ *in* (9.11) *is already recovered. Let us choose an arbitrary 4-vector* $\tilde{\mathbf{H}}_r$ *that is independent of the three columns of* $\mathbf{H}_l$, *and construct a matrix as*

$$\tilde{\mathbf{H}} = [\mathbf{H}_l | \tilde{\mathbf{H}}_r]$$

Then, $\tilde{\mathbf{H}}$ *must be a valid upgrading matrix. i.e.,* $\tilde{\mathbf{M}} = \hat{\mathbf{M}}\tilde{\mathbf{H}}$ *is a valid Euclidean motion matrix, and* $\tilde{\mathbf{S}} = \tilde{\mathbf{H}}^{-1}\hat{\mathbf{S}}$ *corresponds to a valid Euclidean shape matrix.*

Proof Suppose the correct transformation matrix is $\mathbf{H} = [\mathbf{H}_l | \mathbf{H}_r]$, then from

$$\mathbf{S} = \mathbf{H}^{-1}\hat{\mathbf{S}} = \begin{bmatrix} \ell_1\bar{\mathbf{X}}_1, & \ldots, & \ell_n\bar{\mathbf{X}}_n \\ \ell_1, & \ldots, & \ell_n \end{bmatrix} \tag{9.27}$$

we obtain one correct Euclidean structure $[\bar{\mathbf{X}}_1, \ldots, \bar{\mathbf{X}}_n]$ of the object under a certain world coordinate frame by dehomogenizing the shape matrix $\mathbf{S}$. The arbitrary constructed matrix $\tilde{\mathbf{H}} = [\mathbf{H}_l | \tilde{\mathbf{H}}_r]$ and the correct matrix $\mathbf{H}$ are defined up to a 4×4 invertible matrix $\mathbf{G}$ as

$$\mathbf{H} = \tilde{\mathbf{H}}\mathbf{G}, \quad \mathbf{G} = \begin{bmatrix} \mathbf{I}_3 & \mathbf{g} \\ \mathbf{0}^T & s \end{bmatrix} \tag{9.28}$$

where $\mathbf{I}_3$ is a 3×3 identity matrix, $\mathbf{g}$ is a 3-vector, $\mathbf{0}$ is a zero 3-vector, s is a nonzero scalar. Under the transformation matrix $\tilde{\mathbf{H}}$, the motion $\hat{\mathbf{M}}$ and shape $\hat{\mathbf{S}}$ are transformed to

$$\tilde{\mathbf{M}} = \hat{\mathbf{M}}\tilde{\mathbf{H}} = \hat{\mathbf{M}}\mathbf{H}\mathbf{G}^{-1} = \mathbf{M}\begin{bmatrix} \mathbf{I}_3 & -\mathbf{g}/s \\ \mathbf{0}^T & 1/s \end{bmatrix} \tag{9.29}$$

$$\begin{aligned} \tilde{\mathbf{S}} &= \tilde{\mathbf{H}}^{-1}\hat{\mathbf{S}} = (\mathbf{H}\mathbf{G}^{-1})^{-1}\hat{\mathbf{S}} = \mathbf{G}(\mathbf{H}^{-1}\hat{\mathbf{S}}) \\ &= s\begin{bmatrix} \ell_1(\bar{\mathbf{X}}_1 + \mathbf{g})/s & \cdots & \ell_n(\bar{\mathbf{X}}_n + \mathbf{g})/s \\ \ell_1 & \cdots & \ell_n \end{bmatrix} \end{aligned} \tag{9.30}$$

As seen from (9.30), the new shape $\tilde{\mathbf{S}}$ is actually the original structure that undergoes a translation $\mathbf{g}$ and a scale $1/s$, which does not change the Euclidean structure. From (9.29) we have $\tilde{\mathbf{M}}^{(1:3)} = \mathbf{M}^{(1:3)}$, which indicates that the first-three-columns of the new motion matrix (rotation term) do not change. While the last column, which corresponds to translation factor, is modified in accordance with the translation and scale changes of the structure.

Therefore, the constructed matrix $\tilde{\mathbf{H}}$ is a valid transformation matrix that can upgrade the factorization from projective space to the Euclidean space. □

According to Proposition 9.2, the value of $\mathbf{H}_r$ can be set randomly as any 4-vector that is independent of $\mathbf{H}_l$. A practical selection method may be as follows.

Suppose the SVD decomposition of $\mathbf{H}_l$ is

$$\mathbf{H}_l = \mathbf{U}_{4\times 4}\Sigma_{4\times 3}\mathbf{V}_{3\times 3}^T = [\mathbf{u}_1, \mathbf{u}_2, \mathbf{u}_3, \mathbf{u}_4] \begin{bmatrix} \sigma_1 & 0 & 0 \\ 0 & \sigma_2 & 0 \\ 0 & 0 & \sigma_3 \\ 0 & 0 & 0 \end{bmatrix} [\mathbf{v}_1, \mathbf{v}_2, \mathbf{v}_3]^T \tag{9.31}$$

where $\mathbf{U}$ and $\mathbf{V}$ are two orthogonal matrices, Σ is a diagonal of the three singular values. Let us choose an arbitrary value σ_r between the biggest and the smallest singular values of $\mathbf{H}_l$, then we may set

$$\mathbf{H}_r = \sigma_r \mathbf{u}_4, \qquad \mathbf{H} = [\mathbf{H}_l, \mathbf{H}_r] \tag{9.32}$$

The construction guarantees that $\mathbf{H}$ is invertible and has the same condition number as $\mathbf{H}_l$, so that we can obtain a good precision in computing the inverse $\mathbf{H}^{-1}$.

After recovering the Euclidean motion and shape matrices, the intrinsic parameters and pose of the camera associated with each frame can be easily computed as follows.

$$\mu_i = \|\mathbf{M}_{i(3)}^{(1:3)}\| \tag{9.33}$$

$$f_i = \frac{1}{\mu_i}\|\mathbf{M}_{i(1)}^{(1:3)}\| = \frac{1}{\mu_i}\|\mathbf{M}_{i(2)}^{(1:3)}\| \tag{9.34}$$

$$\mathbf{R}_i = \frac{1}{\mu_i}\mathbf{K}_i^{-1}\mathbf{M}_i^{(1:3)}, \qquad \mathbf{T}_i = \frac{1}{\mu_i}\mathbf{K}_i^{-1}\mathbf{M}_i^{(4)} \tag{9.35}$$

where $\mathbf{M}_{i(t)}^{(1:3)}$ denotes the tth row of $\mathbf{M}_i^{(1:3)}$. The result is obtained under quasi-perspective assumption, which is a close approximation to the general perspective projection. The solution may be further optimized to perspective projection by minimizing the image reprojection residuals.

$$J_2 = \min_{(\mathbf{K}_i, \mathbf{R}_i, \mathbf{T}_i, \mu_i, \mathbf{X}_j)} \frac{1}{2}\sum_{i=1}^{m}\sum_{j=1}^{n} |\bar{\mathbf{x}}_{ij} - \hat{\mathbf{x}}_{ij}|^2 \tag{9.36}$$

where $\hat{\mathbf{x}}_{ij}$ denotes the reprojected image point computed via perspective projection (9.1). The minimization process is termed as bundle adjustment, which is usually solved via Levenberg-Marquardt iterations [8].

9.3.2 Algorithm Outline

Given the tracking matrix $\tilde{\mathbf{W}} \in \mathbb{R}^{3m\times n}$ across a sequence with small camera movements. The implementation of the quasi-perspective rigid factorization algorithm is summarized as follows.

1. Balance the tracking matrix via point-wise and image-wise rescalings, as in [16], to improve numerical stability;

9

2. Perform rank-4 SVD factorization on the tracking matrix to obtain a solution of $\hat{\mathbf{M}}$ and $\hat{\mathbf{S}}$;
3. Compute the left part of upgrading matrix $\mathbf{H}_l$ according to (9.19), or (9.21) for negative definite matrix $\mathbf{Q}$;
4. Compute $\mathbf{H}_r$ and $\mathbf{H}$ according to (9.32);
5. Recover the Euclidean motion matrix $\mathbf{M} = \hat{\mathbf{M}}\mathbf{H}$ and shape matrix $\mathbf{S} = \mathbf{H}^{-1}\hat{\mathbf{S}}$;
6. Estimate the camera parameters and pose from (9.33) to (9.35);
7. Optimize the solution via bundle adjustment (9.36).

In the above analysis, as well as in other factorization algorithms, we usually assume one-parameter-camera model as in (9.15) so that we may use this constraint to recover the upgrading matrix $\mathbf{H}$. In real applications, we may take the solution as an initial value and optimize the camera parameters via Kruppa constraints that arise from pairwise images [24].

The essence of quasi-perspective factorization (9.10) is to find a rank-4 approximation $\mathbf{MS}$ of the tracking matrix, i.e. to minimize the Frobenius norm $\|\tilde{\mathbf{W}} - \mathbf{MS}\|_F^2$. Most studies adopt SVD decomposition of $\tilde{\mathbf{W}}$ and truncate it to the desired rank. However, when the tracking matrix is not complete, such as some features are missing in some frames due to occlusions, it is hard to perform SVD decomposition. In case of missing data, we adopt power factorization algorithm [6, 25] to obtain a least squares solution of $\hat{\mathbf{M}}$ and $\hat{\mathbf{S}}$. The solution is then upgraded the solution to Euclidean space according to the proposed scheme.

9.4 Quasi-Perspective Nonrigid Factorization

9.4.1 Problem Formulation

For nonrigid factorization, we follow Bregler's assumption (9.5) to represent a nonrigid shape by weighted combination of k shape bases. Under quasi-perspective projection, the structure is expressed in homogeneous form with nonzero scalars. Let us denote the scale weighted nonrigid structure associated with the ith frame and the lth scale weighted shape basis as

$$\bar{\mathbf{S}}_i = [\ell_1\bar{\mathbf{X}}_1, \ldots, \ell_n\bar{\mathbf{X}}_n], \qquad \mathbf{B}_l = [\ell_1\bar{\mathbf{X}}_1^l, \ldots, \ell_n\bar{\mathbf{X}}_n^l] \tag{9.37}$$

Then from (9.5) we have

$$\bar{\mathbf{X}}_i = \sum_{l=1}^{k} \omega_{il}\bar{\mathbf{X}}_i^l, \quad \forall i = 1, \ldots, n \tag{9.38}$$

Let us multiply a weight scale ℓ_i on both sides as

$$\ell_i \bar{\mathbf{X}}_i = \ell_i \sum_{l=1}^{k} \omega_{il} \bar{\mathbf{X}}_i^l = \sum_{l=1}^{k} \omega_{il} (\ell_i \bar{\mathbf{X}}_i^l) \tag{9.39}$$

then we can immediately have the following result.

$$\mathbf{S}_i = \begin{bmatrix} \bar{\mathbf{S}}_i \\ \ell^T \end{bmatrix} = \begin{bmatrix} \sum_{l=1}^{k} \omega_{il} \mathbf{B}_l \\ \ell^T \end{bmatrix} \tag{9.40}$$

We call (9.40) extended Bregler's assumption to homogeneous case. Under this extension, the quasi-perspective projection of the ith frame can be formulated as

$$\begin{aligned} \tilde{\mathbf{W}}_i &= (\mu_i \mathbf{P}_i)\mathbf{S}_i = [\mu_i \mathbf{P}_i^{(1:3)}, \mu_i \mathbf{P}_i^{(4)}] \begin{bmatrix} \sum_{l=1}^{k} \omega_{il} \mathbf{B}_l \\ \ell^T \end{bmatrix} \\ &= [\omega_{i1}\mu_i \mathbf{P}_i^{(1:3)}, \ldots, \omega_{ik}\mu_i \mathbf{P}_i^{(1:3)}, \mu_i \mathbf{P}_i^{(4)}] \begin{bmatrix} \mathbf{B}_1 \\ \cdots \\ \mathbf{B}_k \\ \ell^T \end{bmatrix} \end{aligned} \tag{9.41}$$

Thus the nonrigid factorization under quasi-perspective projection can be expressed as

$$\tilde{\mathbf{W}}_{3m\times n} = \begin{bmatrix} \omega_{11}\mu_1 \mathbf{P}_1^{(1:3)} & \cdots & \omega_{1k}\mu_1 \mathbf{P}_1^{(1:3)} & \mu_1 \mathbf{P}_1^{(4)} \\ \vdots & \ddots & \vdots & \vdots \\ \omega_{m1}\mu_m \mathbf{P}_m^{(1:3)} & \cdots & \omega_{mk}\mu_m \mathbf{P}_m^{(1:3)} & \mu_m \mathbf{P}_m^{(4)} \end{bmatrix} \begin{bmatrix} \mathbf{B}_1 \\ \vdots \\ \mathbf{B}_k \\ \ell^T \end{bmatrix} \tag{9.42}$$

or represented concisely in matrix form as

$$\tilde{\mathbf{W}}_{3m\times n} = \mathbf{M}_{3m\times(3k+1)} \mathbf{B}_{(3k+1)\times n} \tag{9.43}$$

The factorization expression is similar to (9.7). However, the difficulties in estimating the projective depths are circumvented.

9.4.2 Euclidean Upgrading Matrix

The rank of the tracking matrix is at most $3k+1$, and the factorization is defined up to a transformation matrix $\mathbf{H} \in \mathbb{R}^{(3k+1)\times 3k+1)}$. Suppose the SVD factorization of a tracking matrix with rank constraint is $\tilde{\mathbf{W}} = \hat{\mathbf{M}}\hat{\mathbf{B}}$. Similar to the rigid case, we adopt the metric constraint to compute an upgrading matrix. Let us denote the matrix into $k+1$ parts as

$$\mathbf{H} = [\mathbf{H}_1, \ldots, \mathbf{H}_k | \mathbf{H}_r] \tag{9.44}$$

where $\mathbf{H}_l \in \mathbb{R}^{(3k+1)\times 3}$ $(l = 1, \ldots, k)$ denotes the lth triple columns of $\mathbf{H}$, and $\mathbf{H}_r$ denotes the last column of $\mathbf{H}$. Then we have

$$\hat{\mathbf{M}}_i \mathbf{H}_l = \omega_{il}\mu_i \mathbf{P}_i^{(1:3)} = \omega_{il}\mu_i \mathbf{K}_i \mathbf{R}_i \tag{9.45}$$

$$\hat{\mathbf{M}}_i \mathbf{H}_r = \mu_i \mathbf{P}_i^{(4)} = \mu_i \mathbf{K}_i \mathbf{T}_i \tag{9.46}$$

Similar to (9.15) in rigid case, Let us denote $\mathbf{C}_{ii'} = \hat{\mathbf{M}}_i \mathbf{Q}_l \hat{\mathbf{M}}_{i'}^T$ with $\mathbf{Q}_l = \mathbf{H}_l \mathbf{H}_l^T$, we get

$$\begin{aligned}\mathbf{C}_{ii'} &= \hat{\mathbf{M}}_i \mathbf{Q}_l \hat{\mathbf{M}}_{i'}^T = (\omega_{il}\mu_i \mathbf{K}_i \mathbf{R}_i)(\omega_{i'l}\mu_{i'} \mathbf{K}_{i'} \mathbf{R}_{i'})^T \\ &= \omega_{il}\omega_{i'l}\mu_i\mu_{i'} \mathbf{K}_i (\mathbf{R}_i \mathbf{R}_{i'}) \mathbf{K}_{i'}^T\end{aligned} \tag{9.47}$$

where i and i' $(= 1, \ldots, m)$ correspond to different frame numbers, $l = 1, \ldots, k$ corresponds to different shape bases. Assuming a simplified camera model with only one parameter as $\mathbf{K}_i = \mathrm{diag}(f_i, f_i, 1)$, we have

$$\mathbf{C}_{ii} = \hat{\mathbf{M}}_i \mathbf{Q}_l \hat{\mathbf{M}}_i^T = \omega_{il}^2 \mu_i^2 \begin{bmatrix} f_i^2 & & \\ & f_i^2 & \\ & & 1 \end{bmatrix} \tag{9.48}$$

from which we obtain following four constraints.

$$\begin{cases} f_1(\mathbf{Q}_l) = \mathbf{C}_{ii}(1,2) = 0 \\ f_2(\mathbf{Q}_l) = \mathbf{C}_{ii}(1,3) = 0 \\ f_3(\mathbf{Q}_l) = \mathbf{C}_{ii}(2,3) = 0 \\ f_4(\mathbf{Q}_l) = \mathbf{C}_{ii}(1,1) - \mathbf{C}_{ii}(2,2) = 0 \end{cases} \tag{9.49}$$

The above constraints are similar to (9.16) in rigid case. However, the matrix $\mathbf{Q}_l$ in (9.48) is a $(3k+1) \times (3k+1)$ symmetric matrix. According to Proposition 9.1, $\mathbf{Q}_l$ has $9k$ degrees of freedom, since it can be decomposed into the product of $(3k+1) \times 3$ vertical extended upper triangular matrix. Given m frames, we have $4m$ linear constraints on $\mathbf{Q}_l$. It appears that if we have enough features and frames, the matrix $\mathbf{Q}_l$ can be solved linearly by stacking all constraints in (9.49). Unfortunately, only the rotation constraints may be insufficient when an object deforms at varying speed, since most of these constraints are redundant. Xiao *et al.* [28] proposed a basis constraint to solve this ambiguity.

The main idea of basis constraint is to select k frames that include independent shapes and treat them as a set of bases. Suppose the first k frames are independent of each other, then their corresponding weighting coefficients can be set as

$$\omega_{il} = \begin{cases} 1 & \text{if } i, l = 1, \ldots, k \text{ and } i = l \\ 0 & \text{if } i, l = 1, \ldots, k \text{ and } i \neq l \end{cases} \tag{9.50}$$

From (9.47) we obtain following basis constraint.

$$\mathbf{C}_{ii'} = \begin{bmatrix} 0 & 0 & 0 \\ 0 & 0 & 0 \\ 0 & 0 & 0 \end{bmatrix} \quad \text{if } i = 1, \ldots, k,\ i' = 1, \ldots, m, \text{ and } i \neq l \tag{9.51}$$

Given m images, (9.51) can provide $9m(k-1)$ linear constraints to the matrix $\mathbf{Q}_l$ (some of the constraints are redundant since $\mathbf{Q}_l$ is symmetric). By combining the rotation constraint (9.49) and basis constraint (9.51) together, the matrix $\mathbf{Q}_l$ can be computed linearly. Later, $\mathbf{H}_l, l = 1, \ldots, k$ can be decomposed from $\mathbf{Q}_l$ according to following result.

Result 9.2 *The matrix $\mathbf{Q}_l$ is a $(3k+1) \times (3k+1)$ positive semidefinite symmetric matrix of rank 3. It can be decomposed as $\mathbf{Q} = \mathbf{H}_l \mathbf{H}_l^T$, where $\mathbf{H}_l$ is a $(3k+1) \times 3$ rank 3 matrix. The decomposition can be further written as $\mathbf{Q} = \Lambda_3 \Lambda_3^T$ with Λ_3 being a $(3k+1) \times 3$ vertical extended upper triangular matrix.*

The result can be easily derived from Proposition 9.1. It is easy to verify that the Proposition 9.2 is still valid for nonrigid case. Thus the vector $\mathbf{H}_r$ in (9.44) can be set as an arbitrary $(3k+1)$-vector that is independent of all columns in $\{\mathbf{H}_l\}$, $l = 1, \ldots, k$. After recovering the Euclidean upgrading matrix, the camera parameters, motions, shape bases, weighing coefficients can be easily decomposed from the upgraded motion and shape matrices.

$$\mathbf{M} = \hat{\mathbf{M}}\mathbf{H}, \qquad \mathbf{B} = \mathbf{H}^{-1}\hat{\mathbf{B}} \tag{9.52}$$

9.5 Evaluations on Synthetic Data

9.5.1 Evaluation on Rigid Factorization

During simulations, we randomly generated 200 points within a cube of $20 \times 20 \times 20$ in space and simulated 10 images from these points by perspective projection. The image size is set at 800×800. The camera parameters are set as follows: the focal lengths are set randomly between 900 and 1100, the principal point is set at the image center, and the skew is zero. The rotation angles are set randomly between $\pm 5°$. The X and Y positions of the cameras are set randomly between ± 15, while the Z positions are set evenly from 200 to 220. The imaging condition is close to quasi-perspective assumption.

We add Gaussian white noise to the initially generated 10 images, and vary the noise level from 0 to 3 pixels with steps of 0.5. At each noise level, we reconstruct the 3D structure of the object which is defined up to a similarity transformation with the ground truth. We register reconstructed model with the ground truth and calculate the reconstruction error, which is defined as mean point-wise distance between the reconstructed structure and the ground truth. The mean and standard deviation of the error on 100 independent tests are shown in Fig. 9.1. The proposed algorithm (Quasi) is compared with [13] under affine assumption (Affine) and [5] under perspective projection (Persp). We then perform a bundle adjustment optimization scheme through Levenberg-Marquardt (LM) algorithm [8] to upgrade the solution to perspective projection. It is evident that the proposed method performs much better than that of affine, the optimized solution (Quasi+LM) is very close to perspective projection with optimization (Persp+LM).

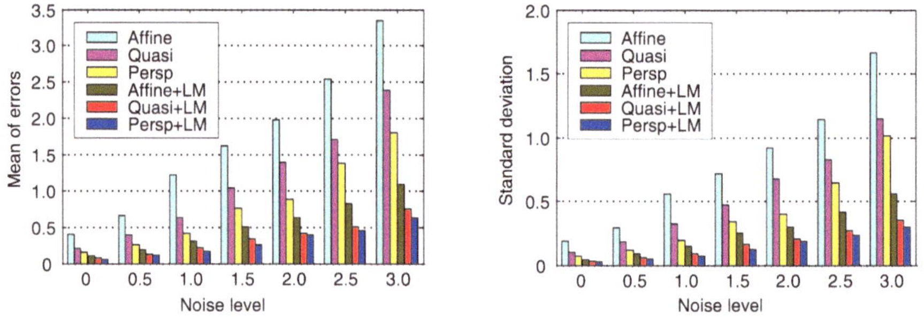

Fig. 9.1 Evaluation on the accuracy of rigid factorization. (**a**) The mean of reconstruction errors by different algorithms at different noise levels; (**b**) The corresponding standard deviation of reconstruction errors

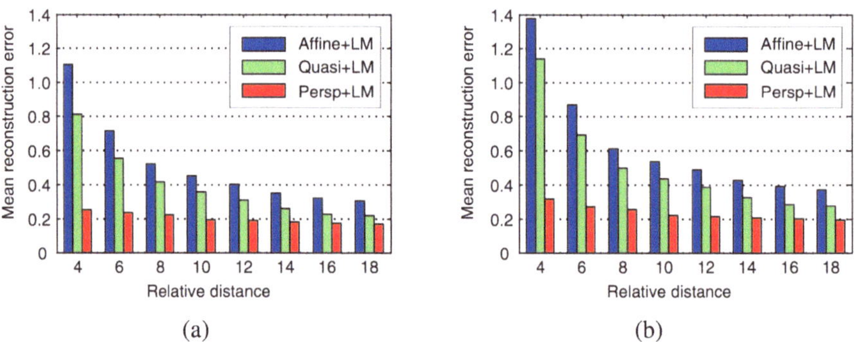

Fig. 9.2 The mean reconstruction error of different projection models with respect to varying relative distance. (**a**) The rotation angle of the camera varies in a range of $\pm 5°$; (**b**) The rotation angle varies in a range of $\pm 20°$

The proposed model is based on the assumption of large relative camera-to-object distance and small camera rotations. We studied the influence of these two factors to different camera models. In the first case, we vary the relative distance from 4 to 18 in steps of 2. At each relative distance, we generated 20 images with the following parameters. The rotation angles are confined between $\pm 5°$, the X and Y positions of the camera are set randomly between ± 15. We recover the structure and compute the reconstruction error for each group of images. The mean reconstruction error by different methods is shown in Fig. 9.2. In the second case, we increase the rotation angles to the range of $\pm 20°$, and retain other camera parameters as in the first case. The mean reconstruction error is given in Fig. 9.2. The results are evaluated on 100 independence tests with 1-pixel Gaussian noise.

Based on experimental evidence we make the following conclusion. (1) The error by quasi-perspective projection is consistently less than that by affine, especially at small relative distances. (2) Both reconstruction errors by affine and quasi-perspective projection increase greatly when the relative distance is less than 6, since both models are based on

Table 9.1 The average computation time of different algorithms

Frame number		5	10	50	100	150	200
	Affine	0.015	0.015	0.031	0.097	0.156	0.219
Time (s)	Quasi	0.015	0.016	0.047	0.156	0.297	0.531
	Persp	0.281	0.547	3.250	6.828	10.58	15.25

large distance assumption. (3) The error at each relative distance increases with the rotation angles, especially at small relative distances, since the projective depths are related to rotation angles. (4) Theoretically, the relative distance and rotation angles have no influence on the result of full perspective projection. However, we see that the error by perspective projection also increases slightly with an increase in rotation angles and a decrease in relative distance. This is because we estimate the projective depths iteratively starting with an affine assumption [5]. The iteration easily gets stuck to local minima due to bad initialization.

We compared the computation time of different factorization algorithms without LM optimization. The program was implemented with Matlab 6.5 on a PC with Intel Pentium 4 3.6 GHz CPU. In this test, we use all 200 feature points and vary the frame number from 5 to 200, so as to generate different data size. The actual computation time (seconds) for different data sets are tabulated in Table 9.1, where computation time for perspective projection is taken for 10 iterations (it usually takes about 30 iterations to compute the projective depths in perspective factorization). Clearly, the computation time required by quasi-perspective factorization is close to that of affine assumption, whereas the perspective factorization is computationally more intensive than other methods.

9.5.2 Evaluation on Nonrigid Factorization

In this test, we generated a synthetic cube with 6 evenly distributed points on each visible edge. There are three sets of moving points on adjacent surfaces of the cube that move at a constant speed as shown in Fig. 9.3, each moving set is composed of 5 points. The cube with moving points can be taken as a nonrigid object with two shape bases. We generated 10 frames with the same camera parameters as in the first test of rigid case. We reconstructed the structure associated with each frame by the proposed method. The result is shown in Fig. 9.3. Results demonstrate that the structure after optimization is visually the same as the ground truth, while the result before optimization is a little bit deformed due to perspective effect.

We compared our method with the nonrigid factorization under affine assumption [27] and that under perspective projection [28]. The mean and standard deviation of the reconstruction errors with respect to different noise levels are shown in Fig. 9.4. It is clear that the proposed method performs significantly better than it does under affine camera model.

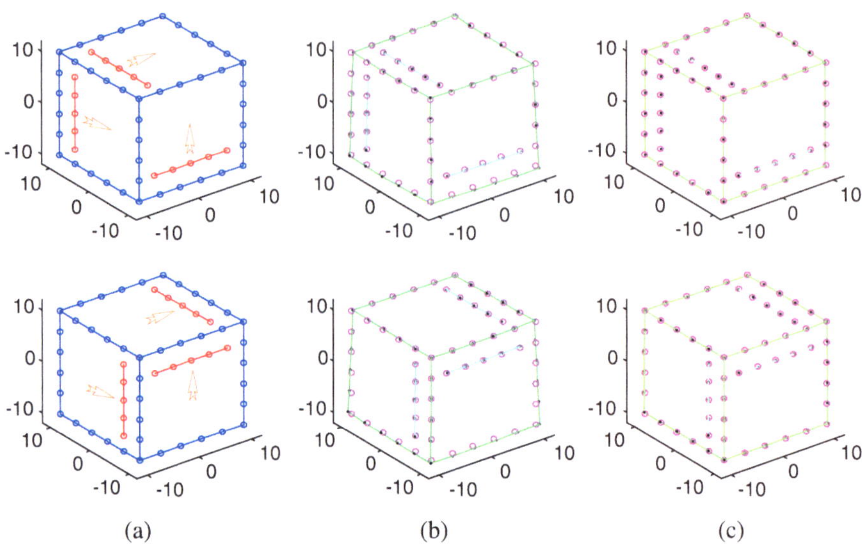

Fig. 9.3 Reconstruction result on nonrigid factorization. (**a**) Two synthetic cubes with moving points in space; (**b**) The quasi-perspective factorization result of the two frames (*in black dots*) superimposed with the ground truth (*in pink circles*); (**c**) The final structures after optimization

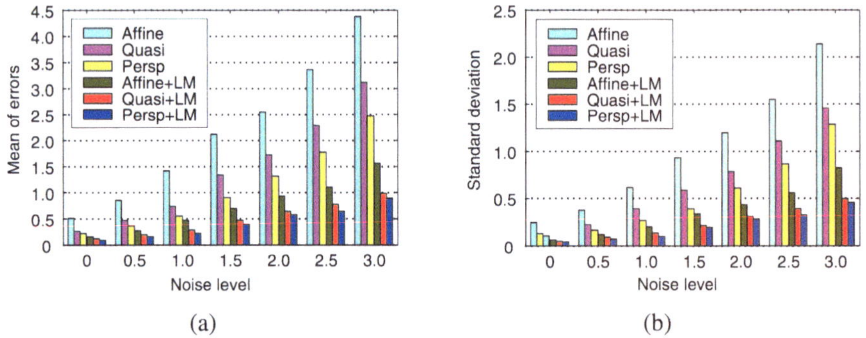

Fig. 9.4 Evaluation on nonrigid factorization. The mean (**a**) and standard deviation (**b**) of the reconstruction errors by different algorithms at different noise levels

9.6 Evaluations on Real Image Sequences

We tested our proposed method on many real sequences, and we report two results in the following.

9.6.1 Test on Fountain Base Sequence

There are 7 images in the fountain base sequence, which were taken at the Sculpture Park of Windsor by Canon Powershot G3 camera. The image resolution is 1024×768. In order to ensure large overlap of the object to be reconstructed, the camera undergoes small movement during image acquisition, hence the quasi-perspective assumption is satisfied for the sequences. We established the initial correspondences by utilizing the technique in [20] and eliminated outliers iteratively as in [12]. Totally 4218 reliable features were tracked across the sequence as shown in Fig. 9.5. We recovered 3D structure of the object and camera motions by utilizing the proposed algorithm, as well as some previous methods. Figure 9.5 shows the reconstructed VRML model with texture mapping and the corresponding triangulated wireframes from different viewpoints. The model looks realistic and most details are correctly recovered by the proposed method.

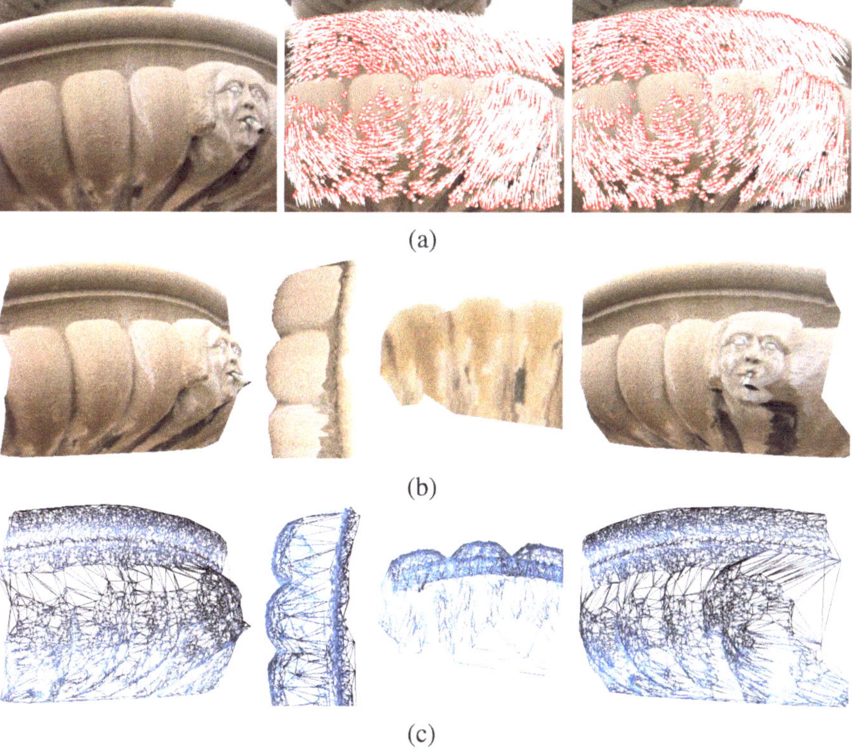

Fig. 9.5 Reconstruction result of fountain base sequence. (**a**) Three images from the sequence, where the tracked features with relative disparities are overlaid to the second and the third images; (**b**) The reconstructed VRML model of the scene shown from different viewpoints with texture mapping; (**c**) The corresponding triangulated wireframe of the reconstructed model

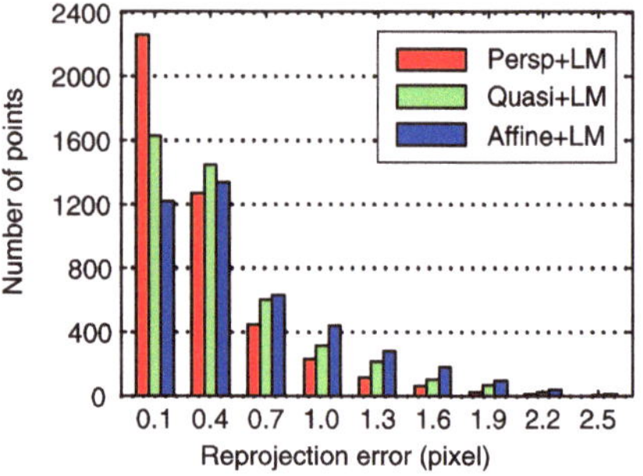

Fig. 9.6 The histogram distributions of the reprojection errors by different algorithms in the test of fountain base sequence

Table 9.2 Camera parameters of the first four frames and reprojection errors in real sequence test

Method	f_1	f_2	f_3	f_4	Mean	STD	$\mathbf{E}_{rep}$
Quasi+LM	2140.5	2143.6	2139.4	2142.8	0.418	0.285	2.473
Affine+LM	2153.4	2155.7	2151.2	2153.1	0.629	0.439	3.189
Persp+LM	2131.7	2135.3	2131.2	2134.5	0.240	0.168	1.962

In order to compare the algorithms quantitatively, we reproject the reconstructed 3D structure back to the images and calculate the reprojection errors, i.e. distances between detected and reprojected image points. Figure 9.6 shows the histogram distributions of the errors using 9 bins. The corresponding mean ('Mean') and standard deviation ('STD') of the errors are listed in Table 9.2. We see that the reprojection error by our proposed model is much smaller than that under affine assumption.

9.6.2 Test on Franck Sequence

The Franck face sequence was downloaded from the European working group on face and gesture recognition. We selected 60 frames with various facial expressions for the test. The image resolution is 720 × 576, and there are 68 tracked feature across the sequence. Figure 9.7 shows the reconstructed models of two frames utilizing the proposed nonrigid factorization method. Different facial expressions are correctly recovered. As a comparison, the relative reprojection error $\mathbf{E}_{rep}$ generated from different methods are listed in Table 9.2. All tests illustrate that the accuracy by the proposed method is fairly close to that of full perspective projection, and considerably better than affine assumption.

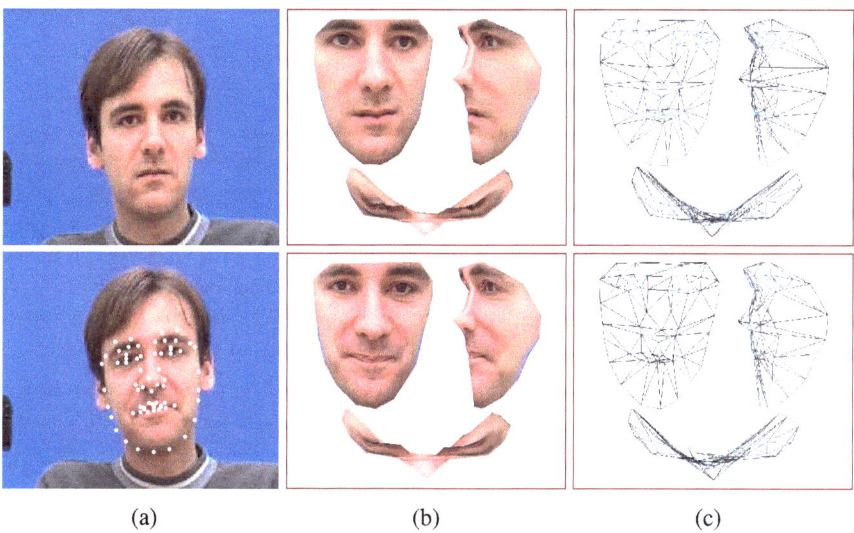

(a) (b) (c)

Fig. 9.7 Reconstruction of different facial expressions in Franck face sequence. (**a**) Two frames from the sequence with the 68 tracked features overlaid to the lower frame; (**b**) Front, side, and top views of the reconstructed VRML models with texture mapping; (**c**) The corresponding triangulated wireframe of the reconstructed model

9.7 Closure Remarks

9.7.1 Conclusion

In this chapter, we proposed a quasi-perspective factorization algorithm for both rigid and nonrigid objects and elaborated the computation details of Euclidean upgrading matrix. The proposed method avoids the difficulties associated with computation of projective depths under perspective factorization. It is computationally simple with better accuracy than affine approximation. The proposed model is suitable for structure and motion factorization of short sequences with small camera motions. Experiments demonstrates improvements of our algorithm over existing techniques. For a long image sequences taken around an object, the assumption of small camera movements is violated. However, we can simply divide the sequence into several subsequences with small movements, then register and merge the result of each subsequence to reconstruct the structure of the whole object.

9

9.7.2 Review Questions

1. *Rigid factorization.* Present the extended Cholesky decomposition. Show how Cholesky decomposition can be used to recover $\mathbf{H}_l$ in the upgrading matrix. When the matrix $\mathbf{Q}$ is negative definite, how to estimate $\mathbf{H}_l$? Give a cost function to upgrade the quasi-perspective solution to full perspective projection?
2. *Nonrigid factorization.* Derive the nonrigid factorization expression under quasi-perspective projection and list its differences with affine and perspective nonrigid factorization. Prove that $\mathbf{H}_r$ in the nonrigid upgrading matrix can be set freely as an arbitrary $(3k+1)$-vector that is independent of all columns in $\{\mathbf{H}_l\}$.

References

1. Brand, M.: A direct method for 3D factorization of nonrigid motion observed in 2D. In: Proc. of IEEE Conference on Computer Vision and Pattern Recognition, vol. 2, pp. 122–128 (2005)
2. Bregler, C., Hertzmann, A., Biermann, H.: Recovering non-rigid 3D shape from image streams. In: Proc. of IEEE Conference on Computer Vision and Pattern Recognition, vol. 2, pp. 690–696 (2000)
3. Costeira, J., Kanade, T.: A multibody factorization method for independent moving objects. Int. J. Comput. Vis. **29**(3), 159–179 (1998)
4. Del Bue, A., Lladó, X., de Agapito, L.: Non-rigid metric shape and motion recovery from uncalibrated images using priors. In: Proc. of IEEE Conference on Computer Vision and Pattern Recognition, vol. 1, pp. 1191–1198 (2006)
5. Han, M., Kanade, T.: Creating 3D models with uncalibrated cameras. In: Proc. of IEEE Computer Society Workshop on the Application of Computer Vision (2000)
6. Hartley, R., Schaffalizky, F.: Powerfactorization: 3D reconstruction with missing or uncertain data. In: Proc. of Australia-Japan Advanced Workshop on Computer Vision (2003)
7. Hartley, R., Vidal, R.: Perspective nonrigid shape and motion recovery. In: Proc. of European Conference on Computer Vision, vol. 1, pp. 276–289. Springer, Berlin (2008)
8. Hartley, R.I., Zisserman, A.: Multiple View Geometry in Computer Vision, 2nd edn. Cambridge University Press, Cambridge (2004). ISBN: 0521540518
9. Li, T., Kallem, V., Singaraju, D., Vidal, R.: Projective factorization of multiple rigid-body motions. In: Proc. of IEEE Conference on Computer Vision and Pattern Recognition (2007)
10. Mahamud, S., Hebert, M.: Iterative projective reconstruction from multiple views. In: Proc. of IEEE Conference on Computer Vision and Pattern Recognition, vol. 2, pp. 430–437 (2000)
11. Oliensis, J., Hartley, R.: Iterative extensions of the Sturm/Triggs algorithm: Convergence and nonconvergence. IEEE Trans. Pattern Anal. Mach. Intell. **29**(12), 2217–2233 (2007)
12. Torr, P.H.S., Zisserman, A., Maybank, S.J.: Robust detection of degenerate configurations while estimating the fundamental matrix. Comput. Vis. Image Underst. **71**(3), 312–333 (1998)
13. Poelman, C., Kanade, T.: A paraperspective factorization method for shape and motion recovery. IEEE Trans. Pattern Anal. Mach. Intell. **19**(3), 206–218 (1997)
14. Quan, L.: Self-calibration of an affine camera from multiple views. Int. J. Comput. Vis. **19**(1), 93–105 (1996)
15. Rabaud, V., Belongie, S.: Re-thinking non-rigid structure from motion. In: Proc. of IEEE Conference on Computer Vision and Pattern Recognition (2008)
16. Sturm, P.F., Triggs, B.: A factorization based algorithm for multi-image projective structure and motion. In: Proc. of European Conference on Computer Vision, vol. 2, pp. 709–720 (1996)

17. Tomasi, C., Kanade, T.: Shape and motion from image streams under orthography: A factorization method. Int. J. Comput. Vis. **9**(2), 137–154 (1992)
18. Torresani, L., Hertzmann, A., Bregler, C.: Nonrigid structure-from-motion: Estimating shape and motion with hierarchical priors. IEEE Trans. Pattern Anal. Mach. Intell. **30**(5), 878–892 (2008)
19. Triggs, B.: Factorization methods for projective structure and motion. In: Proc. of the IEEE Conference on Computer Vision and Pattern Recognition, pp. 845–851 (1996)
20. Wang, G.: A hybrid system for feature matching based on SIFT and epipolar constraints. Tech. Rep. Department of ECE, University of Windsor (2006)
21. Wang, G., Tsui, H.T., Hu, Z.: Structure and motion of nonrigid object under perspective projection. Pattern Recogn. Lett. **28**(4), 507–515 (2007)
22. Wang, G., Tsui, H.T., Wu, J.: Rotation constrained power factorization for structure from motion of nonrigid objects. Pattern Recogn. Lett. **29**(1), 72–80 (2008)
23. Wang, G., Wu, J.: Quasi-perspective projection with applications to 3D factorization from uncalibrated image sequences. In: Proc. of IEEE Conference on Computer Vision and Pattern Recognition, pp. 1–8 (2008)
24. Wang, G., Wu, J.: Perspective 3D Euclidean reconstruction with varying camera parameters. IEEE Trans. Circuits Syst. Video Technol. **19**(12), 1793–1803 (2009)
25. Wang, G., Wu, J.: Stratification approach for 3-D Euclidean reconstruction of nonrigid objects from uncalibrated image sequences. IEEE Trans. Syst. Man Cybern., Part B **38**(1), 90–101 (2008)
26. Wang, G., Wu, J.: Quasi-perspective projection model: Theory and application to structure and motion factorization from uncalibrated image sequences. Int. J. Comput. Vis. **87**(3), 213–234 (2010)
27. Xiao, J., Chai, J., Kanade, T.: A closed-form solution to non-rigid shape and motion recovery. Int. J. Comput. Vis. **67**(2), 233–246 (2006)
28. Xiao, J., Kanade, T.: Uncalibrated perspective reconstruction of deformable structures. In: Proc. of the International Conference on Computer Vision, vol. 2, pp. 1075–1082 (2005)
29. Yan, J., Pollefeys, M.: A factorization-based approach to articulated motion recovery. In: Proc. of IEEE Conference on Computer Vision and Pattern Recognition, vol. 2, pp. 815–821 (2005)
30. Yan, J., Pollefeys, M.: A factorization-based approach for articulated nonrigid shape, motion and kinematic chain recovery from video. IEEE Trans. Pattern Anal. Mach. Intell. **30**(5), 865–877 (2008)

Projective Geometry for Computer Vision 

Projective geometry is all geometry.

Arthur Cayley (1821–1895)

We are familiar with the concept and measurements of Euclidean geometry, which is a good approximation to the properties of a general physical space. However, when we consider the imaging process of a camera, the Euclidean geometry becomes insufficient since parallelism, lengths, and angles are no longer preserved in images. In this appendix, we will briefly survey some basic concepts and properties of projective geometry which are extensively used in computer vision. For further information, readers may refer to [1, 3, 7].

Euclidean geometry is actually a subset of the projective geometry, which is more general and least restrictive in the hierarchy of fundamental geometries. Just like Euclidean geometry, projective geometry exists in any number of dimensions, such as a line in one-dimensional projective space, denoted as $\mathbb{P}^1$, corresponds to 1D Euclidean space $\mathbb{R}^1$; the projective plane in $\mathbb{P}^2$ is analogous to 2D Euclidean plane; the three-dimensional projective space $\mathbb{P}^3$ is related to 3D Euclidean space.

A.1 2D Projective Geometry

A.1.1 Points and Lines

In Euclidean space $\mathbb{R}^2$, a point can be denoted as $\bar{\mathbf{x}} = [x, y]^T$, a line passing through the point can be represented as

$$l_1x + l_2y + l_3 = 0 \tag{A.1}$$

If we multiply the same nonzero scalar w on both sides of (A.1), we have

$$l_1xw + l_2yw + l_3w = 0 \tag{A.2}$$

G. Wang, Q.M.J. Wu, *Guide to Three Dimensional Structure and Motion Factorization*, Advances in Pattern Recognition,
DOI 10.1007/978-0-85729-046-5, © Springer-Verlag London Limited 2011

Clearly, (A.1) and (A.2) represent the same line. Let $\mathbf{x} = [xw, yw, w]^T, \mathbf{l} = [l_1, l_2, l_3]^T$, then the line (A.2) is represented as

$$\mathbf{x}^T \mathbf{l} = \mathbf{l}^T \mathbf{x} = 0 \tag{A.3}$$

where the line is represented by the vector $\mathbf{l}$, and any point on the line is denoted by $\mathbf{x}$. We call the 3-vector $\mathbf{x}$ the homogeneous coordinates of a point in $\mathbb{P}^2$, which represents the same point as inhomogeneous coordinates $\bar{\mathbf{x}} = [xw/w, yw/w]^T = [x, y]^T$. Similarly, we call $\mathbf{l}$ the homogeneous representation of the line, since for any nonzero scalar k, $\mathbf{l}$ and $k\mathbf{l}$ represent the same line.

From (A.3), we find that there is actually no difference between the representation of a line and the representation of a point. This is known as the duality principal. Given two lines $\mathbf{l} = [l_1, l_2, l_3]^T$ and $\mathbf{l}' = [l_1', l_2', l_3']^T$, their intersection defines a point that can be computed from

$$\mathbf{x} = \mathbf{l} \times \mathbf{l}' = [\mathbf{l}]_\times \mathbf{l}' \tag{A.4}$$

where '$\times$' denotes the cross product of two vectors,

$$[\mathbf{l}]_\times = \begin{bmatrix} 0 & -t_3 & t_2 \\ t_3 & 0 & -t_1 \\ -t_2 & t_1 & 0 \end{bmatrix}$$

denotes the antisymmetric matrix of vector $\mathbf{l}$. Similarly, a line passing through two points $\mathbf{x}$ and $\mathbf{x}'$ can be computed from

$$\mathbf{l} = \mathbf{x} \times \mathbf{x}' = [\mathbf{x}]_\times \mathbf{x}' = [\mathbf{x}']_\times \mathbf{x} \tag{A.5}$$

Any point with homogeneous coordinates $\mathbf{x} = [x, y, 0]^T$ corresponds to a point at infinity, or ideal point. Whereas its corresponding inhomogeneous point $\bar{\mathbf{x}} = [x/0, y/0]^T$ makes no sense. In space plane, all ideal points can be written as $[x, y, 0]^T$. The set of these points lies on a single line $\mathbf{l}_\infty$, which is called the line at infinity. From (A.3), it is easy to obtain the coordinates of the line at infinity $\mathbf{l}_\infty = [0, 0, 1]^T$.

A.1.2 Conics and Duel Conics

In Euclidean plane, the equation of a conic in inhomogeneous coordinates is written as

$$ax^2 + bxy + cy^2 + dx + ey + f = 0 \tag{A.6}$$

If we adopt homogeneous coordinates and denote any point on the conic by $\mathbf{x} = [x_1, x_2, x_3]^T$, then the conic (A.6) can be written as the following quadratic homogeneous expression.

$$ax_1^2 + bx_1x_2 + cx_2^2 + dx_1x_3 + ex_2x_3 + fx_3 = 0 \tag{A.7}$$

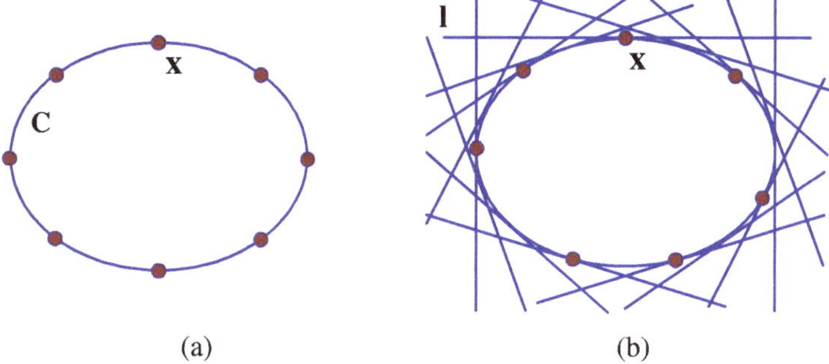

Fig. A.1 A point conic (**a**) and its dual line conic (**b**). $\mathbf{x}$ is a point on the conic $\mathbf{x}^T\mathbf{C}\mathbf{x}=0$, $\mathbf{l}$ is a line tangent to $\mathbf{C}$ at point $\mathbf{x}$ which satisfies $\mathbf{l}^T\mathbf{C}^*\mathbf{l}=0$

or in a matrix form as

$$\mathbf{x}^T\mathbf{C}\mathbf{x}=[x_1,x_2,x_3]\begin{bmatrix} a & b/2 & d/2 \\ b/2 & c & e/2 \\ d/2 & e/2 & f \end{bmatrix}\begin{bmatrix} x_1 \\ x_2 \\ x_3 \end{bmatrix}=0 \tag{A.8}$$

where $\mathbf{C}$ is the conic coefficient matrix which is symmetric. A conic has five degrees of freedom since multiplying $\mathbf{C}$ by any nonzero scalar does not affect the above equation. Therefore, five points in $\mathbb{P}^2$ at a general position (no three points are collinear) can uniquely determine a conic. Generally, a conic matrix $\mathbf{C}$ is of full rank. In degenerate cases, it may degenerate to two lines when $rank(\mathbf{C})=2$, or one repeated line when $rank(\mathbf{C})=1$.

The conic defined in (A.8) is defined by points in $\mathbb{P}^2$, which is usually termed as point conic. According to duality principal, we can obtain the dual line conic as

$$\mathbf{l}^T\mathbf{C}^*\mathbf{l}=0 \tag{A.9}$$

where the notation $\mathbf{C}^*$ stands for the adjoint matrix of $\mathbf{C}$. The dual conic is also called conic envelope, as shown in Fig. A.1, which is formulated by lines tangent to $\mathbf{C}$. For conics (A.8) and (A.9), we have the following Results.

Result A.1 *The line* $\mathbf{l}$ *tangent to the non-degenerate conic* $\mathbf{C}$ *at point* $\mathbf{x}$ *is given by* $\mathbf{l}=\mathbf{C}\mathbf{x}$. *In duality, the tangent point* $\mathbf{x}$ *to the non-degenerate line conic* $\mathbf{C}^*$ *at line* $\mathbf{l}$ *is given by* $\mathbf{x}=\mathbf{C}^*\mathbf{l}$.

Result A.2 *For non-degenerate conic* $\mathbf{C}$ *and its duality* $\mathbf{C}^*$, *we have* $\mathbf{C}^*=\mathbf{C}^{-1}$, *and* $(\mathbf{C}^*)^*=\mathbf{C}$. *The line conics may degenerate to two points when* $rank(\mathbf{C}^*)=2$, *or one repeated point when* $rank(\mathbf{C}^*)=1$, *and* $(\mathbf{C}^*)^*\neq\mathbf{C}$ *in degenerate cases.*

Any point $\mathbf{x}$ and a conic $\mathbf{C}$ define a line $\mathbf{l}=\mathbf{C}\mathbf{x}$, as shown in Fig. A.2. Then $\mathbf{x}$ and $\mathbf{l}$ forms a pole-polar relationship. The point $\mathbf{x}$ is called the pole of line $\mathbf{l}$ with respect to conic $\mathbf{C}$, and the line $\mathbf{l}$ is called the polar of point $\mathbf{x}$ with respect to conic $\mathbf{C}$. It is easy to verify the following Result.

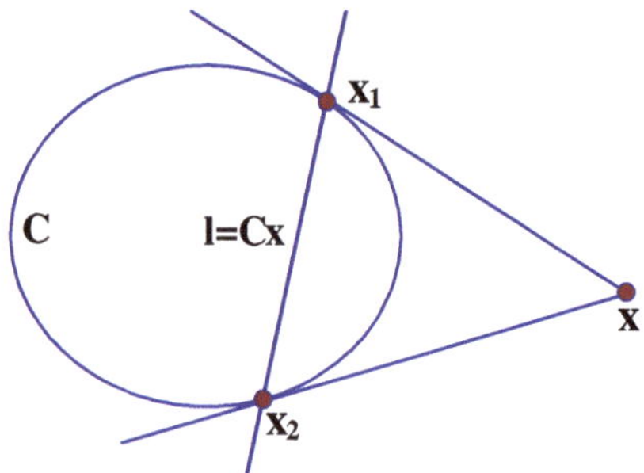

Fig. A.2 The pole-polar relationship. The line $\mathbf{l} = \mathbf{C}\mathbf{x}$ is the polar of point $\mathbf{x}$ with respect to conic $\mathbf{C}$, and the point $\mathbf{x} = \mathbf{C}^{-1}\mathbf{l}$ is the pole of $\mathbf{l}$ with respect to conic $\mathbf{C}$

Result A.3 *The polar line $\mathbf{l} = \mathbf{C}\mathbf{x}$ intersects the conic $\mathbf{C}$ at two points $\mathbf{x}_1$ and $\mathbf{x}_2$, then the two lines $\mathbf{l}_1 = \mathbf{x} \times \mathbf{x}_1$ and $\mathbf{l}_2 = \mathbf{x} \times \mathbf{x}_2$ are tangent to the conic $\mathbf{C}$. If the point $\mathbf{x}$ is on the conic, then the polar is the tangent line to $\mathbf{C}$ at $\mathbf{x}$.*

Result A.4 *Any two points $\mathbf{x}$ and $\mathbf{y}$ satisfying $\mathbf{x}^T\mathbf{C}\mathbf{y} = 0$ are called conjugate points with respect to $\mathbf{C}$. The set of all conjugate points of $\mathbf{x}$ forms the polar line $\mathbf{l}$. If $\mathbf{x}$ is on the polar of $\mathbf{x}'$, then $\mathbf{x}'$ is also on the polar of $\mathbf{x}$ since $\mathbf{x}^T\mathbf{C}\mathbf{x}' = \mathbf{x}'^T\mathbf{C}\mathbf{x} = 0$. In duality, two lines $\mathbf{l}$ and $\mathbf{l}'$ are conjugate with respect to $\mathbf{C}$ if $\mathbf{l}^T\mathbf{C}\mathbf{l}' = 0$.*

Result A.5 *There are a pair of conjugate ideal points*

$$\mathbf{i} = \begin{bmatrix} 1 \\ i \\ 0 \end{bmatrix}, \qquad \mathbf{j} = \begin{bmatrix} 1 \\ -i \\ 0 \end{bmatrix}$$

on the line at infinity $\mathbf{l}_\infty$. We call $\mathbf{i}$ and $\mathbf{j}$ the canonical forms of circular points. Essentially, the circular points are the intersection of any circle with the line at infinity. Thus three additional points can uniquely determine a circle, which is equivalent to the fact that five general points can uniquely determine a general conic. The dual of the circular point forms a degenerated line conic given by $\mathbf{C}^*_\infty = \mathbf{i}\mathbf{j}^T + \mathbf{j}\mathbf{i}^T$.

A.1.3 2D Projective Transformation

Two dimensional projective transformation is an invertible linear mapping $\mathbf{H} : \mathbb{P}^2 \to \mathbb{P}^2$ which is a 3×3 matrix. The transformation is also known as projectivity, or homography. The mapping of a point $\mathbf{x} = [x_1, x_2, x_3]^T$ can be written as

$$\begin{bmatrix} x_1' \\ x_2' \\ x_3' \end{bmatrix} = \begin{bmatrix} h_{11} & h_{12} & h_{13} \\ h_{21} & h_{22} & h_{23} \\ h_{31} & h_{32} & h_{33} \end{bmatrix} \begin{bmatrix} x_1 \\ x_2 \\ x_3 \end{bmatrix} \tag{A.10}$$

or more briefly as $\mathbf{x}' = \mathbf{H}\mathbf{x}$. This is a homogeneous transformation which is defined up to scale. Thus there are only 8 degrees of freedom on $\mathbf{H}$. Four pairs of corresponding points can uniquely determine the transformation if no three points are collinear. The transformation (A.10) is defined by points.

Result A.6 *Under a point transformation* $\mathbf{x}' = \mathbf{H}\mathbf{x}$, *a line* $\mathbf{l}$ *is transformed to* $\mathbf{l}'$ *via*

$$\mathbf{l}' = \mathbf{H}^{-1}\mathbf{l} \tag{A.11}$$

A conic $\mathbf{C}$ *is transformed to* $\mathbf{C}'$ *via*

$$\mathbf{C}' = \mathbf{H}^{-T}\mathbf{C}\mathbf{H}^{-1} \tag{A.12}$$

and a dual conic $\mathbf{C}^*$ *is transformed to* $\mathbf{C}^{*\prime}$ *via*

$$\mathbf{C}^{*\prime} = \mathbf{H}\mathbf{C}^*\mathbf{H}^T \tag{A.13}$$

All projective transformations form a group which is called projective linear group. There are some specializations or subgroups of the transformation, such as affine group, Euclidean group, and oriented Euclidean group. Different transformations have different geometric invariance and properties. For example, length and area are invariant under Euclidean transformation; parallelism and line at infinity are invariant under affine transformation; general projective transformation preserves concurrency, collinearity, and cross ratio.

A.2 3D Projective Geometry

A.2.1 Points, Lines, and Planes

In 3D space $\mathbb{P}^3$, the homogeneous coordinates of a point is represented by a 4-vector as $\mathbf{X} = [x_1, x_2, x_3, x_4]^T$, which is defined up to a scale since $\mathbf{X}$ and $s\mathbf{X}$ ($s \neq 0$) represent the same point. The corresponding inhomogeneous coordinates is $\bar{\mathbf{X}} = [x, y, z]^T = [x_1/x_4, x_2/x_4, x_3/x_4]^T$. When $X_4 = 0$, $\mathbf{X}$ represents a point at infinity.

A plane in 3D space $\mathbb{P}^3$ can be formulated as

$$\Pi\mathbf{X} = \pi_1 x_1 + \pi_2 x_2 + \pi_3 x_3 + \pi_4 x_4 = 0 \tag{A.14}$$

where $\mathbf{X} = [x_1, x_2, x_3, x_4]^T$ is the homogeneous representation of a point on the plane. The 4-vector $\Pi = [\pi_1, \pi_2, \pi_3, \pi_4]^T$ is called the homogeneous coordinates of the plane. When $\Pi = [0, 0, 0, 1]^T$, the solution of (A.14) is the set of all points at infinity. In this case, the plane is named as plane at infinity, denoted as Π_∞. For any finite plane $\Pi \neq \Pi_\infty$, if we use inhomogeneous point coordinates, the plane equation in Euclidean geometry can be written as

$$\mathbf{n}^T\bar{\mathbf{X}} + d = 0 \tag{A.15}$$

where $\mathbf{n} = [\pi_1, \pi_2, \pi_3]^T$ is called the plane normal; $d = \pi_4$, and $d/\|\mathbf{n}\|$ is the distance of the plane from origin.

Result A.7 *Two distinct planes intersect in a unique line. Two planes are parallel if and only if their intersection is a line at infinity. A line is parallel to a plane if and only if their intersection is a point at infinity.*

Result A.8 *Three non-collinear points $\mathbf{X}_1$, $\mathbf{X}_2$, and $\mathbf{X}_3$ uniquely define a plane, which can be obtained from the 1-dimensional right null-space of the 3×4 matrix $\mathbf{A} = [\mathbf{X}_1, \mathbf{X}_2, \mathbf{X}_3]^T$ since $\mathbf{A}\Pi = 0$.*

Result A.9 *As a dual to Result A.8, three non-collinear planes Π_1, Π_2, and Π_3 uniquely define a point, which can be obtained from the 1-dimensional right null-space of the 3×4 matrix $\mathbf{A}^* = [\Pi_1, \Pi_2, \Pi_3]^T$ since $(\mathbf{A}^*)\mathbf{X} = 0$.*

A line in $\mathbb{P}^3$ is defined by the join of two points or the intersection of two planes, which has 4 degrees of freedom in 3-space. Suppose $\mathbf{X}_1$ and $\mathbf{X}_2$ are two non-coincident space points, then the line joining these points can be defined by the span of the two points with the following 2×4 matrix

$$\mathbf{A} = \begin{bmatrix} \mathbf{X}_1^T \\ \mathbf{X}_2^T \end{bmatrix} \tag{A.16}$$

It is evident that the pencil of points $\mathbf{X} = \{\mathbf{X}(\alpha, \beta) = \alpha\mathbf{X}_1 + \beta\mathbf{X}_2\}$ is a line joining the two points, which is called the line generated by the span of $\mathbf{A}^T$. The 2-dimensional right null-space of $\mathbf{A}$ is a pencil of planes with the line as axis.

Similarly, suppose we have two planes Π_1 and Π_2. A dual representation of a line can be generated from the span of the 2×4 matrix

$$\mathbf{A}^* = \begin{bmatrix} \Pi_1^T \\ \Pi_2^T \end{bmatrix} \tag{A.17}$$

Clearly, the span of $\mathbf{A}^{*T}$ is a pencil of planes $\Pi = \{\Pi(\alpha, \beta) = \alpha\Pi_1 + \beta\Pi_2\}$ with the line as axis. The 2-dimensional right null-space of $\mathbf{A}^*$ is the pencil of points on the line. There are still other popular representations of line by virtue of Plücker matrices and Plücker line coordinates.

A.2.2 Projective Transformation and Quadrics

Similar to the 2D projective transformation (A.10), the transformation is $\mathbb{P}^3$ can be described by a 4×4 matrix $\mathbf{H}$ as

$$\mathbf{X}' = \mathbf{H}\mathbf{X} \tag{A.18}$$

The transformation $\mathbf{H}$ is a homogeneous matrix defined up to a scale, thus it has only 15 degrees of freedom. The hierarchy of 3D transformation matrix includes subgroups of Euclidean, affine, and projective transformations. Each subgroup has its special form and invariant properties.

A quadric is a surface in $\mathbb{P}^3$ defined by the following equation.

$$\mathbf{X}^T\mathbf{Q}\mathbf{X} = 0 \tag{A.19}$$

where $\mathbf{Q}$ is a 4×4 symmetric matrix. The quadric has some similar properties as the conic in $\mathbb{P}^2$.

Result A.10 *A quadric has 9 degrees of freedom since it is defined up to a scale. Thus 9 points in general position can define a quadric. If the matrix $\mathbf{Q}$ is singular, the quadric is degenerate which may be defined by fewer points.*

Result A.11 *A quadric defines a polarity between a point and a plane. The plane $\Pi = \mathbf{Q}\mathbf{X}$ is the polar plane of $\mathbf{X}$ with respect to $\mathbf{Q}$. A plane intersects a quadric at a conic.*

Result A.12 *Under the transformation $\mathbf{X}' = \mathbf{H}\mathbf{X}$, a quadric transforms as*

$$\mathbf{Q}' = \mathbf{H}^{-T}\mathbf{Q}\mathbf{H}^{-1} \tag{A.20}$$

and a dual quadric $\mathbf{Q}^*$ transforms as

$$\mathbf{Q}^{*\prime} = \mathbf{H}\mathbf{Q}^*\mathbf{H}^T \tag{A.21}$$

Matrix Decomposition

B

Mathematics is the door and key to the sciences.

Roger Bacon (1214–1294)

In linear algebra, a matrix decomposition is a factorization of a matrix into some canonical form. There are many different classes of matrix decompositions. In this appendix, we will introduce some common decomposition methods used in this book, such as singular value decomposition, RQ decomposition, and Cholesky decomposition, etc. Please refer to [2, 9] for more details.

B.1 Singular Value Decomposition

Singular value decomposition (SVD) is one of the most useful decompositions in numerical computations such as optimization, least-squares, etc. In this book, it is widely used as a basic mathematical tool for the factorization algorithm.

Suppose $\mathbf{A}$ is an $m \times n$ real matrix. The singular value decomposition of $\mathbf{A}$ is of the form

$$\mathbf{A} = \mathbf{U}\Sigma\mathbf{V}^T \tag{B.1}$$

where $\mathbf{U}$ is an $m \times m$ orthogonal matrix whose columns are the eigenvectors of $\mathbf{A}\mathbf{A}^T$, $\mathbf{V}$ is an $n \times n$ orthogonal matrix whose columns are the eigenvectors of $\mathbf{A}^T\mathbf{A}$, and Σ is an $m \times n$ diagonal matrix with nonnegative real numbers on the diagonal. The decomposition is conventionally carried out in such a way that the diagonal entries σ_i are arranged in a descending order. The diagonal entries of Σ are known as the singular values of $\mathbf{A}$.

G. Wang, Q.M.J. Wu, *Guide to Three Dimensional Structure and Motion Factorization*,
Advances in Pattern Recognition,
DOI 10.1007/978-0-85729-046-5,

B

B.1.1 Properties of SVD Decomposition

SVD decomposition reveals many intrinsic properties of matrix **A** and is numerically stable for computation. Suppose the singular values of **A** are

$$\sigma_1 \geq \sigma_2 \geq \cdots \geq \sigma_r > \sigma_{r+1} = \cdots = 0$$

Then from (B.1) we infer the following statements about SVD.

1. $\mathbf{A} = \sum_{i=1}^{r} \sigma_i \mathbf{u}_i \mathbf{v}_i^T = \mathbf{U}_r \Sigma_r \mathbf{V}_r^T$,
   where $\mathbf{U}_r = [\mathbf{u}_1, \ldots, \mathbf{u}_r]$, $\mathbf{V}_r = [\mathbf{v}_1, \ldots, \mathbf{v}_r]$, $\Sigma_r = \text{diag}\,(\sigma_1, \ldots, \sigma_r)$;
2. $rank(\mathbf{A}) = rank(\Sigma) = r$.
   The rank of **A** and Σ equal to r;
3. $range(\mathbf{A}) = span\{\mathbf{u}_1, \ldots, \mathbf{u}_r\}$.
   The column space of **A** is spanned by the first r columns of **U**;
4. $null(\mathbf{A}) = span\{\mathbf{v}_{r+1}, \ldots, \mathbf{v}_n\}$.
   The null space of **A** is spanned by the last $n - r$ columns of **V**;
5. $range(\mathbf{A}^T) = span\{\mathbf{v}_1, \ldots, \mathbf{v}_r\}$.
   The row space of **A** is spanned by the first r columns of **V**;
6. $null(\mathbf{A}^T) = span\{\mathbf{u}_{r+1}, \ldots, \mathbf{u}_m\}$.
   The null space of **A** is spanned by the last $m - r$ columns of **U**;
7. $\|\mathbf{A}\|_F = (\sigma_1^2 + \cdots + \sigma_r^2)^{\frac{1}{2}}$;
8. $\|\mathbf{A}\|_2 = \sigma_1$.

From SVD decomposition (B.1), we have

$$\mathbf{A}\mathbf{A}^T = \mathbf{U}\Sigma\mathbf{V}^T\mathbf{V}\Sigma^T\mathbf{U}^T = \mathbf{U}\Sigma^2\mathbf{U}^T \tag{B.2}$$

Therefore, $\sigma_i^2, i = 1, \ldots, m$ are eigenvalues of $\mathbf{A}\mathbf{A}^T$, and the columns of **U** are the corresponding eigenvectors. Similarly, from

$$\mathbf{A}^T\mathbf{A} = \mathbf{V}\Sigma^T\mathbf{U}^T\mathbf{U}\Sigma\mathbf{V}^T = \mathbf{V}\Sigma^2\mathbf{V}^T \tag{B.3}$$

we know that $\sigma_i^2, i = 1, \ldots, n$ are eigenvalues of $\mathbf{A}^T\mathbf{A}$, and the columns of **V** are the corresponding eigenvectors.

The result of SVD decomposition can be used to measure the dependency between columns of a matrix. The measure is termed as condition number and defined by

$$cond(\mathbf{A}) = \frac{\sigma_{max}}{\sigma_{min}} \tag{B.4}$$

where σ_{max} and σ_{min} denote the largest and smallest singular values of **A**. Note that $cond(\mathbf{A}) \geq 1$. If the condition number is close to 1, then the columns of **A** are independent. Large condition number means the columns of **A** are nearly dependent. If **A** is singular, $\sigma_{min} = 0$, and $cond(\mathbf{A}) = \infty$. The condition number plays an important role in the numerical solution of linear systems since it measures the noise sensitivity of the systems.

B.1.2
Low-Rank Matrix Approximation

The fact that $rank(\mathbf{A}) = rank(\Sigma)$ tells us that we can determine the rank of matrix $\mathbf{A}$ by counting the nonzero entries in Σ. In some practical applications, such as the structure and motion factorization algorithm, we need to solve the problem of approximating a matrix with another matrix which has a specific low rank. For example, an $m \times n$ matrix is supposed to have rank r, but due to noise data, its rank is greater than r. This will show up in Σ. When we look at Σ we may find that $\sigma_{r+1}, \sigma_{r+2}, \ldots$ are much smaller than σ_r and very close to zero. In this case, we can obtain the best rank-r approximation of the matrix by modifying Σ.

If $rank(\mathbf{A}) > r$ and we want to approximate it with a rank-r matrix $\hat{\mathbf{A}}$, the approximation is based on minimizing the Frobenius norm of the difference between $\mathbf{A}$ and $\tilde{\mathbf{A}}$

$$J = \min_{\tilde{\mathbf{A}}} \|\mathbf{A} - \hat{\mathbf{A}}\|_F \tag{B.5}$$

subject to the constraint that $rank(\tilde{\mathbf{A}}) = r$. It turns out that the solution can be simply given by singular value decomposition. Suppose the SVD factorization of $\mathbf{A}$ is

$$\mathbf{A} = \mathbf{U}\Sigma\mathbf{V}^T \tag{B.6}$$

Let us construct a diagonal matrix $\tilde{\Sigma}$ which is the same matrix as Σ except that it contains only the r largest singular values and the rest are replaced by zero. Then, the rank-r approximation is given by

$$\tilde{\mathbf{A}} = \mathbf{U}\tilde{\Sigma}\mathbf{V}^T \tag{B.7}$$

This is known as the Eckar-Young theorem in linear algebra. Here is a short proof of the theorem.

Proof Since the Frobenius norm is unitarily invariant, we have the equivalent cost function of (B.5) as

$$J = \min_{\tilde{\mathbf{A}}} \|\mathbf{U}^T\mathbf{A}\mathbf{V} - \mathbf{U}^T\hat{\mathbf{A}}\mathbf{V}\|_F = \min_{\tilde{\mathbf{A}}} \|\Sigma - \tilde{\Sigma}\|_F = \min_{\tilde{\sigma}_i} \left(\sum_{i=1}^{n} (\sigma_i - \tilde{\sigma}_i)^2 \right)^{\frac{1}{2}} \tag{B.8}$$

where $\tilde{\sigma}_i$ is the singular value of $\tilde{\mathbf{A}}$, and $\tilde{\sigma}_{r+1} = \cdots = \tilde{\sigma}_n = 0$. Thus, the cost function (B.8) is converted to

$$J = \min_{\tilde{\sigma}_i} \left(\sum_{i=1}^{r} (\sigma_i - \tilde{\sigma}_i)^2 + \sum_{i=r+1}^{n} \sigma_i^2 \right)^{\frac{1}{2}} = \left(\sum_{i=r+1}^{n} \sigma_i^2 \right)^{\frac{1}{2}} \tag{B.9}$$

Therefore, $\tilde{\mathbf{A}}$ is the best rank-r approximation of $\mathbf{A}$ in the Frobenius norm sense when $\tilde{\sigma}_i = \sigma_i$ for $i = 1, \ldots, r$, and $\tilde{\sigma}_{r+1} = \cdots = \tilde{\sigma}_n = 0$. □

B

B.2
QR and RQ Decompositions

An orthogonal matrix $\mathbf{Q}$ is a square matrix whose transpose equals to its inverse as

$$\mathbf{Q}^T\mathbf{Q} = \mathbf{Q}\mathbf{Q}^T = \mathbf{I} \tag{B.10}$$

where $\mathbf{I}$ is the identity matrix. Taking determinants on (B.10) leads to $\det(\mathbf{Q}) = \pm 1$. If $\det(\mathbf{Q}) = 1$, $\mathbf{Q}$ is a rotation matrix. Otherwise, it is called a reflection matrix.

QR decomposition is to decompose a matrix into an orthogonal matrix and an upper-triangular matrix. Any real square matrix $\mathbf{A}$ can be decomposed as

$$\mathbf{A} = \mathbf{QR} \tag{B.11}$$

where $\mathbf{Q}$ is an orthogonal matrix, and $\mathbf{R}$ is an upper-triangular matrix, also called right-triangular matrix. QR decomposition is often used to solve the linear least squares problem. Analogously, we can define QL, RQ, and LQ decompositions, with $\mathbf{L}$ being a left-triangular matrix. The decomposition is usually performed by means of the Gram-Schmidt process, Householder transformation, or Givens rotations.

In computer vision, we are particularly interested in the RQ decomposition of a 3×3 real matrix. Using this decomposition, we can factorize the projection matrix into a camera parameter matrix and a rotation matrix. The process can be easily realized by Givens rotations.

The 3-dimensional Givens rotations can be given by

$$\mathbf{G}_x = \begin{bmatrix} 1 & & \\ & c & -s \\ & s & c \end{bmatrix}, \quad \mathbf{G}_y = \begin{bmatrix} c & & s \\ & 1 & \\ -s & & c \end{bmatrix}, \quad \mathbf{G}_z = \begin{bmatrix} c & -s & \\ s & c & \\ & & 1 \end{bmatrix} \tag{B.12}$$

where $c = \cos\theta$ and $s = \sin\theta$ with θ a rotation angle. Multiply matrix $\mathbf{A}$ on the right by a Givens rotation (B.12), we can zero an element in the subdiagonal of the matrix, forming the triangular matrix $\mathbf{R}$. The concatenation of all the Givens rotations forms the orthogonal matrix $\mathbf{Q}$.

For example, the RQ decomposition of a matrix $\mathbf{A} = [a_{ij}]_{3\times 3}$ can be performed as follows.

1. Zero the subdiagonal element a_{21} by multiplying $\mathbf{G}_z$;
2. Zero the subdiagonal element a_{31} by multiplying $\mathbf{G}_y$. The process does not change a_{21};
3. Zero the subdiagonal element a_{32} by multiplying $\mathbf{G}_x$. The process does not change a_{21} and a_{31};
4. Formulate the decomposition $\mathbf{A} = \mathbf{RQ}$, where the upper-triangular matrix is given by $\mathbf{R} = \mathbf{A}\mathbf{G}_z\mathbf{G}_y\mathbf{G}_x$, and the rotation matrix is given by $\mathbf{Q} = \mathbf{G}_x^T\mathbf{G}_y^T\mathbf{G}_z^T$.

B.3 Symmetric and Skew-Symmetric Matrix

A symmetric matrix $\mathbf{A}$ is a square matrix that is equal to its transpose. i.e. $\mathbf{A} = \mathbf{A}^T$. A skew-symmetric (also called antisymmetric) matrix $\mathbf{A}$ is a square matrix whose transpose is its negative. i.e. $\mathbf{A} = -\mathbf{A}^T$.

Some properties of symmetric and skew-symmetric matrices are listed bellow.

1. Every diagonal matrix is symmetric, since all off-diagonal entries are zero. Similarly, each diagonal element of a skew-symmetric matrix must be zero.
2. Any square matrix can be expressed as the sum of symmetric and skew-symmetric parts.

$$\mathbf{A} = \frac{1}{2}(\mathbf{A} + \mathbf{A}^T) + \frac{1}{2}(\mathbf{A} - \mathbf{A}^T) \tag{B.13}$$

   where $\mathbf{A} + \mathbf{A}^T$ is symmetric, $\mathbf{A} - \mathbf{A}^T$ is skew-symmetric.
3. Let $\mathbf{A}$ be an $n \times n$ skew-symmetric matrix. The determinant of $\mathbf{A}$ satisfies

$$\det(\mathbf{A}) = \det(\mathbf{A}^T) = \det(-\mathbf{A}) = (-1)^n \det(\mathbf{A}) \tag{B.14}$$

   Thus, $\det(\mathbf{A}) = 0$ when n is odd.
4. Every real symmetric matrix $\mathbf{A}$ can be diagonalized. If $\mathbf{A}$ is a real $n \times n$ symmetric matrix, its eigen decomposition has the following simple form.

$$\mathbf{A} = \mathbf{U}\mathbf{D}\mathbf{U}^T = \mathbf{U}\begin{bmatrix} \sigma_1^2 & & \\ & \ddots & \\ & & \sigma_n^2 \end{bmatrix}\mathbf{U}^T \tag{B.15}$$

   where $\mathbf{U}$ is an $n \times n$ orthogonal matrix whose columns are eigenvectors of $\mathbf{A}$, $\mathbf{D}$ is an $n \times n$ orthogonal matrix with the eigenvalues of $\mathbf{A}$ as the diagonal elements. This is equivalent to the matrix equation

$$\mathbf{A}\mathbf{U} = \mathbf{U}\mathbf{D} \tag{B.16}$$

B.3.1 Cross Product

A 3-vector $\mathbf{a} = [a_1, a_2, a_3]^T$ defines a corresponding 3×3 skew-symmetric matrix as follows.

$$[\mathbf{a}]_\times = \begin{bmatrix} 0 & -a_3 & a_2 \\ a_3 & 0 & -a_1 \\ -a_2 & a_1 & 0 \end{bmatrix} \tag{B.17}$$

B

The matrix $[\mathbf{a}]_\times$ is singular and $\mathbf{a}$ is one of its null-vector. Thus a 3×3 skew-symmetric matrix is defined up to a scale by its null-vector. Reversely, any 3×3 skew-symmetric matrix can be written as $[\mathbf{a}]_\times$.

The cross product of two 3-vectors $\mathbf{a}=[a_1,a_2,a_3]^T$ and $\mathbf{b}=[b_1,b_2,b_3]^T$ is given by

$$\mathbf{a}\times\mathbf{b}=[a_2b_3-a_3b_2,a_3b_1-a_1b_3,a_1b_2-a_2b_1]^T \tag{B.18}$$

It is easy to verify that the cross product is related with the skew-symmetric matrix as

$$\mathbf{a}\times\mathbf{b}=[\mathbf{a}]_\times\mathbf{b}=\left[\mathbf{a}^T[\mathbf{b}]_\times\right]^T \tag{B.19}$$

Suppose $\mathbf{A}$ is any 3×3 matrix, $\mathbf{a}$ and $\mathbf{b}$ are two 3-vectors. Then we have

$$(\mathbf{Aa})\times(\mathbf{Ab})=\mathbf{A}^*[\mathbf{a}\times\mathbf{b}] \tag{B.20}$$

where $\mathbf{A}^*$ is the adjoint of $\mathbf{A}$. Specifically, when $\mathbf{A}$ is invertible, equation (B.20) can be written as

$$(\mathbf{Aa})\times(\mathbf{Ab})=\det(\mathbf{A})\mathbf{A}^{-T}[\mathbf{a}\times\mathbf{b}] \tag{B.21}$$

B.3.2 Cholesky Decomposition

A symmetric matrix $\mathbf{A}$ is positive definite if, for any vector $\mathbf{x}$, the product $\mathbf{x}^T\mathbf{A}\mathbf{x}$ is positive. A positive definite symmetric matrix can be uniquely decomposed as

$$\mathbf{A}=\mathbf{K}\mathbf{K}^T$$

where $\mathbf{K}$ is an upper-triangular real matrix with positive diagonal entries. Such a decomposition is called Cholesky decomposition, which is of great importance in camera calibration.

Proof Following the eigen decomposition (B.15), we have

$$\mathbf{A}=\mathbf{UDU}^T=\left(\mathbf{UD}^{\frac{1}{2}}\right)\left(\mathbf{UD}^{\frac{1}{2}}\right)^T=\mathbf{VV}^T \tag{B.22}$$

Perform RQ decomposition on $\mathbf{V}$, we have $\mathbf{V}=\mathbf{KQ}$, where $\mathbf{K}$ is an upper-triangular real matrix, $\mathbf{Q}$ is an orthogonal matrix. Then the decomposition (B.22) can be written as

$$\mathbf{A}=(\mathbf{KQ})(\mathbf{KQ})^T=\mathbf{KK}^T \tag{B.23}$$

In (B.23), the diagonal entries of $\mathbf{K}$ may not be positive. Suppose the sign of the ith diagonal entry of $\mathbf{K}$ is $\mathrm{sign}(k_{ii})$. Then, we can make them positive by multiplying a diagonal transformation matrix $\mathbf{H}=\mathrm{diag}(\mathrm{sign}(k_{11}),\ldots,\mathrm{sign}(k_{nn}))$ to $\mathbf{K}$. This will not change the form of (B.23).

The decomposition is unique. Otherwise, suppose there are two such decompositions as

$$\mathbf{A} = \mathbf{K}_1\mathbf{K}_1^T = \mathbf{K}_2\mathbf{K}_2^T$$

from which we have

$$\mathbf{K}_2^{-1}\mathbf{K}_1 = \mathbf{K}_2^T\mathbf{K}_1^{-T} = (\mathbf{K}_2^{-1}\mathbf{K}_1)^{-T}$$

where $\mathbf{K}_2^{-1}\mathbf{K}_1$ is a diagonal matrix. Thus, $\mathbf{K}_2^{-1}\mathbf{K}_1 = \mathbf{I}$, and $\mathbf{K}_2 = \mathbf{K}_1$. □

B.3.3 Extended Cholesky Decomposition

Cholesky Decomposition can only deal with positive definite symmetric matrix. In some applications, such as recovering the upgrading matrix in Chap. 9, the matrix $\mathbf{A}$ is a positive semidefinite symmetric matrix. In this case, we employ the following extended Cholesky decomposition.

Suppose $\mathbf{A}$ is a $n \times n$ positive semidefinite symmetric matrix of rank k $(k < n)$. Then it can be decomposed as $\mathbf{A} = \mathbf{H}_k\mathbf{H}_k^T$, where $\mathbf{H}_k$ is a $n \times k$ matrix of rank k. Furthermore, the decomposition can be written as $\mathbf{A} = \Lambda_k\Lambda_k^T$ with Λ_k a $n \times k$ vertical extended upper triangular matrix. The degree-of-freedom of the matrix $\mathbf{A}$ is $nk - \frac{1}{2}k(k-1)$, which is the number of unknowns in Λ_k.

Proof Since $\mathbf{A}$ is a $n \times n$ positive semidefinite symmetric matrix of rank k, it can be decomposed by SVD as

$$\mathbf{A} = \mathbf{U}\Sigma\mathbf{U}^T = \mathbf{U}\begin{bmatrix} \sigma_1 & & & & & \\ & \ddots & & & & \\ & & \sigma_k & & & \\ & & & 0 & & \\ & & & & \ddots & \\ & & & & & 0 \end{bmatrix}\mathbf{U}^T \tag{B.24}$$

where $\mathbf{U}$ is a $n \times n$ orthogonal matrix, Σ is a diagonal matrix with σ_i the singular value of $\mathbf{A}$. Consequently we get

$$\mathbf{H}_k = \mathbf{U}^{(1:k)}\begin{bmatrix} \sqrt{\sigma_1} & & \\ & \ddots & \\ & & \sqrt{\sigma_k} \end{bmatrix} = \begin{bmatrix} \mathbf{H}_{ku} \\ \mathbf{H}_{kl} \end{bmatrix} \tag{B.25}$$

such that $\mathbf{A} = \mathbf{H}_k\mathbf{H}_k^T$, where $\mathbf{U}^{(1:k)}$ denotes first k columns of $\mathbf{U}$, $\mathbf{H}_{ku}$ denotes upper $(n-k) \times k$ submatrix of $\mathbf{H}_k$, and $\mathbf{H}_{kl}$ denotes lower $k \times k$ submatrix of $\mathbf{H}_k$. By applying RQ-decomposition to $\mathbf{H}_{kl}$, we have $\mathbf{H}_{kl} = \Lambda_{kl}\mathbf{O}_k$, where Λ_{kl} is an upper triangular matrix, $\mathbf{O}_k$ is an orthogonal matrix.

Let us denote $\Lambda_{ku} = \mathbf{H}_{ku}\mathbf{O}_k^T$ and construct a $n \times k$ vertical extended upper triangular matrix

$$\Lambda_k = \begin{bmatrix} \Lambda_{ku} \\ \Lambda_{kl} \end{bmatrix} \tag{B.26}$$

Then we have

$$\mathbf{H}_k = \Lambda_k \mathbf{O}_k, \qquad \mathbf{A} = \mathbf{H}_k \mathbf{H}_k^T = (\Lambda_k \mathbf{O}_k)(\Lambda_k \mathbf{O}_k)^T = \Lambda_k \Lambda_k^T \tag{B.27}$$

It is easy to verify that the matrix $\mathbf{A}$ has $nk - \frac{1}{2}k(k-1)$ degrees of freedom, which is just the number of unknowns in Λ_k. □

Numerical Computation Method C

What we know is not much. What we do not know is immense.

Pierre-Simon Laplace (1749–1827)

In this appendix, we will introduce two widely used numerical computation methods. One is least squares for linear systems, the other is iterative estimation method for nonlinear systems. Please refer to [2, 8, 9] for detailed study.

C.1 Linear Least Squares

The method of least squares is a standard approach to the approximate solution of an overdetermined system of equations. "Least squares" means that the overall solution minimizes the sum of the squares of the residuals. Assuming normal distribution of the errors, least squares solution produces a maximum likelihood estimation of the parameters.

Depending on whether or not the residuals are linear in all unknowns, least squares problems fall into two categories: linear least squares and nonlinear least squares. Let us consider a linear system of the form

$$\mathbf{A}\mathbf{x} = \mathbf{b} \tag{C.1}$$

where $\mathbf{A} \in \mathbb{R}^{m\times n}$ is the data matrix, $\mathbf{x} \in \mathbb{R}^n$ is the parameter vector, and $\mathbf{b} \in \mathbb{R}^m$ is the observation vector. The condition for uniqueness of the solution is closely related to the rank of matrix $\mathbf{A}$.

The rank of $\mathbf{A}$ is the maximum number of linearly independent column vectors of $\mathbf{A}$. Thus the rank can never exceed the number of columns. On the other hand, since $rank(\mathbf{A}) = rank(\mathbf{A}^T)$, we also know that the rank of $\mathbf{A}$ also equals to the maximum number of linearly independent rows of $\mathbf{A}$. Therefore, $rank(\mathbf{A}) \leq \min(m, n)$. We say that $\mathbf{A}$ is full rank if $rank(\mathbf{A}) = \min(m, n)$.

In the system (C.1), when $m < n$, there are more unknowns than equations. Then $\mathbf{A}$ does not have full column rank, the solution is not unique. When $m = n$ and $\mathbf{A}$ is of full rank,

G. Wang, Q.M.J. Wu, *Guide to Three Dimensional Structure and Motion Factorization*, Advances in Pattern Recognition,
DOI 10.1007/978-0-85729-046-5, © Springer-Verlag London Limited 2011

there will be a unique solution. When $m > n$, typically the system has no exact solution unless $\mathbf{b}$ happens to lie in the span of the columns of $\mathbf{A}$. The goal of least squares fitting is to find $\mathbf{x} \in \mathbb{R}^n$ to minimize

$$J_1 = \min_{\mathbf{x}} \|\mathbf{A}\mathbf{x} - \mathbf{b}\|_2 \tag{C.2}$$

C.1.1 Full Rank System

We consider the case when $m > n$ and $\mathbf{A}$ is of full rank. Three methods will be presented in the following section.

Normal equation. Suppose $\mathbf{x}$ is the solution to the least squares problem (C.2). Then $\mathbf{A}\mathbf{x}$ is the closest point to $\mathbf{b}$, and $\mathbf{A}\mathbf{x} - \mathbf{b}$ must be a vector orthogonal to the column space of $\mathbf{A}$. Thus we have $\mathbf{A}^T(\mathbf{A}\mathbf{x} - \mathbf{b}) = 0$, which leads to

$$(\mathbf{A}^T\mathbf{A})\mathbf{x} = \mathbf{A}^T\mathbf{b} \tag{C.3}$$

We call (C.3) the normal equations. Since $rank(\mathbf{A}) = n$, $\mathbf{A}^T\mathbf{A}$ is an $n \times n$ positive definite symmetric matrix. Thus, (C.3) has unique solution which can be given by

$$\mathbf{x} = (\mathbf{A}^T\mathbf{A})^{-1}\mathbf{A}^T\mathbf{b} \tag{C.4}$$

The normal equations can also be solved by Cholesky decomposition as follows.

1. Let $\mathbf{C} = \mathbf{A}^T\mathbf{A}$ and $\mathbf{d} = \mathbf{A}^T\mathbf{b}$;
2. Perform Cholesky decomposition $\mathbf{C} = \mathbf{K}\mathbf{K}^T$;
3. Solve $\mathbf{K}\mathbf{y} = \mathbf{d}$ by forward substitution;
4. Solve $\mathbf{K}^T\mathbf{x} = \mathbf{y}$ by backward substitution.

The computation cost of the algorithm is about $mn^2 + \frac{1}{3}n^3$.

QR decomposition. Suppose the QR decomposition of $\mathbf{A}$ is $\mathbf{QR}$, then we have

$$\mathbf{A}^T\mathbf{A} = (\mathbf{QR})^T(\mathbf{QR}) = \mathbf{R}^T\mathbf{R}$$

and the system (C.3) is converted to

$$(\mathbf{R}^T\mathbf{R})\mathbf{x} = \mathbf{R}^T\mathbf{Q}^T\mathbf{b} \tag{C.5}$$

Since $\mathbf{R}$ is invertible, (C.5) can be simplified to

$$\mathbf{R}\mathbf{x} = \mathbf{Q}^T\mathbf{b} = \mathbf{d} \tag{C.6}$$

Thus, $\mathbf{x}$ can be easily solved by backward substitution from (C.6). The computation cost of the algorithm is about $2mn^2$.

SVD decomposition. Since $rank(\mathbf{A}) = n$, the SVD decomposition of $\mathbf{A}$ has the following form

$$\mathbf{A} = \mathbf{U}\Sigma\mathbf{V}^T = \mathbf{U}\begin{bmatrix}\Sigma_n & \\ & \mathbf{0}\end{bmatrix}\mathbf{V}^T \tag{C.7}$$

We define the pseudo-inverse of $\mathbf{A}$ as

$$\mathbf{A}^+ = \mathbf{V}\Sigma^+\mathbf{U}^T = \mathbf{V}\begin{bmatrix}\Sigma_n^{-1} & \\ & \mathbf{0}\end{bmatrix}\mathbf{U}^T \tag{C.8}$$

Therefore, the least squares solution is given by

$$\mathbf{x} = \mathbf{A}^+\mathbf{b} = \mathbf{V}\begin{bmatrix}\Sigma_n^{-1} & \\ & \mathbf{0}\end{bmatrix}\mathbf{U}^T\mathbf{b} = \sum_{i=1}^{n}\frac{\mathbf{u}_i^T\mathbf{b}}{\sigma_i}\mathbf{v}_i \tag{C.9}$$

C.1.2 Deficient Rank System

In some cases, the column vectors of $\mathbf{A}$ are not independent of each other. Suppose $rank(\mathbf{A}) = r < n$, then the solution is not unique. There will be an $(n-r)$-parameter family of solutions. The above discussed normal equations and QR decomposition are no longer valid for deficient rank system. However, we can still use SVD decomposition. Let

$$\mathbf{A} = \mathbf{U}\Sigma\mathbf{V}^T = \mathbf{U}\begin{bmatrix}\Sigma_r & \\ & \mathbf{0}\end{bmatrix}\mathbf{V}^T \tag{C.10}$$

Then one solution can be obtained from

$$\mathbf{x}_0 = \mathbf{A}^+\mathbf{b} = \mathbf{V}\begin{bmatrix}\Sigma_r^{-1} & \\ & \mathbf{0}\end{bmatrix}\mathbf{U}^T\mathbf{b} = \sum_{i=1}^{r}\frac{\mathbf{u}_i^T\mathbf{b}}{\sigma_i}\mathbf{v}_i \tag{C.11}$$

The general solution is given by

$$\mathbf{x} = \mathbf{x}_0 + \mu_{r+1}\mathbf{v}_{r+1} + \cdots + \mu_n\mathbf{v}_n = \sum_{i=1}^{r}\frac{\mathbf{u}_i^T\mathbf{b}}{\sigma_i}\mathbf{v}_i + \sum_{j=r+1}^{n}\mu_j\mathbf{v}_j \tag{C.12}$$

which is a $(n-r)$-parameter family parameterized by μ_j.

C.1.2.1 Homogeneous System

We consider a homogeneous system of equation as

$$\mathbf{A}\mathbf{x} = 0 \tag{C.13}$$

C

where $\mathbf{A}$ is an $m \times n$ matrix. Suppose the system is over-determined, i.e. $m > n$. We are seeking a nonzero solution to the system. Obviously, if $\mathbf{x}$ is a solution, $k\mathbf{x}$ is also a solution for any scalar k. Thus we apply a constraint $\|\mathbf{x}\| = 1$. The least squares problem can be written as

$$J_2 = \min_{\mathbf{x}} \|\mathbf{Ax}\|_2, \quad s.t. \|\mathbf{x}\| = 1 \tag{C.14}$$

Case 1. If $rank(\mathbf{A}) = r < n$, a general solution of (C.14) is given by

$$\mathbf{x} = \sum_{i=1}^{n-r} \alpha_i \mathbf{v}_{r+i}, \quad s.t. \sum_{i=1}^{n-r} \alpha_i^2 = 1$$

where $\mathbf{v}_{r+1}, \ldots, \mathbf{v}_n$ are $n - r$ linear independent eigenvectors of $\mathbf{A}^T\mathbf{A}$.

Proof The problem (C.14) is equivalent to

$$J_2 = \min_{\mathbf{x}} \|\mathbf{A}^T\mathbf{Ax}\|_2, \quad s.t. \|\mathbf{x}\| = 1 \tag{C.15}$$

Since $rank(\mathbf{A}^T\mathbf{A}) = rank(\mathbf{A}) = r$, the SVD decomposition of $\mathbf{A}^T\mathbf{A}$ can be written as follows.

$$\mathbf{A}^T\mathbf{A} = \mathbf{V}\,\mathrm{diag}(\lambda_1, \ldots, \lambda_r, 0, \ldots, 0)\mathbf{V}^T \tag{C.16}$$

It is easy to verify from (C.16) that $\mathbf{v}_{r+1}, \ldots, \mathbf{v}_n$ are $n - r$ linearly independent mutually orthogonal solutions of (C.15). □

Case 2. If $rank(\mathbf{A}) = n - 1$, system (C.14) has a unique solution which is the eigenvector corresponding to the zero eigenvalue of $\mathbf{A}^T\mathbf{A}$. For noise data, $\mathbf{A}$ may be of full rank. Then the least squares solution can be given by the eigenvector that corresponds to the smallest eigenvalue of $\mathbf{A}^T\mathbf{A}$.

In practice, we simply perform SVD decomposition $\mathbf{A} = \mathbf{U}\Sigma\mathbf{V}^T$, then the last column of $\mathbf{V}$ is exactly the solution.

C.2 Nonlinear Estimation Methods

In the following part, we will briefly introduce the bundle adjustment method and two commonly used iterative algorithms for nonlinear parameter estimation.

C.2.1 Bundle Adjustment

Bundle adjustment is always used as the last refining step of every feature-based reconstruction algorithm. The objective is to produce jointly optimal estimation of the 3D structure and projection parameters. The problem of structure from motion is defined as: given a set of tracking data $\mathbf{x}_{ij}$, $i = 1, \ldots, m$, $j = 1, \ldots, n$, which are the projection of n 3D points over m views, we want to recover the camera projection parameters $\mathbf{P}_i$ and the coordinates of the space points $\mathbf{X}_j$ following the imaging process $\mathbf{x}_{ij} = \mathbf{P}_i\mathbf{X}_j$.

Suppose we already have a set of solution to the problem. However, due to image noise and model error, the projection equations are not satisfied exactly. We want to refine the estimated projection matrices $\hat{\mathbf{P}}_i$ and 3D points $\hat{\mathbf{X}}_j$ so that their projection $\hat{\mathbf{x}}_{ij} = \hat{\mathbf{P}}_i\hat{\mathbf{X}}_j$ is more close to the measurement $\mathbf{x}_{ij}$. The process can be realized by minimizing the image residues (i.e., the geometric distance between the detected image point $\mathbf{x}_{ij}$ and the reprojected point $\hat{\mathbf{x}}_{ij}$) as follows [3, 10].

$$\min_{\mathbf{P}_i, \mathbf{X}_j} \sum_{i,j} d(\mathbf{x}_{ij} - \hat{\mathbf{P}}_i\hat{\mathbf{X}}_j)^2 \tag{C.17}$$

If the image error is zero-mean Gaussian, then bundle adjustment produces a maximum likelihood estimation. The cost function (C.17) involves a large number of nonlinear equations and parameters. Thus, the minimization is achieved using nonlinear least-squares algorithms, such as Newton iteration and Levenberg-Marquardt iteration (LM). The modified sparse LM algorithm has proven to be one of the most successful algorithms due to its efficiency and the ability to converge quickly from a wide range of initial guesses.

More generally, an unconstrained nonlinear estimation problem can be usually denoted as

$$\min_{\mathbf{x}} f(\mathbf{x}) \tag{C.18}$$

where $\mathbf{x} \in \mathbb{R}^n$ is a parameter vector, $f(\mathbf{x}) \in \mathbb{R}^m$ is a nonlinear convex function. We want to find the parameter vector $\hat{\mathbf{x}}$ such that (C.18) is minimized. The problem of nonlinear optimization is usually solved via a iterative algorithm as follows.

1. The algorithm starts from an initial estimation $\mathbf{x}_0$;
2. Compute the next search direction and increments to construct the next point $\mathbf{x}_{i+1} = \mathbf{x}_i + \Delta_i$;
3. Perform termination test for minimization.

The main difference between different algorithms lies in the construction method of shift vector Δ_i, which is a key factor of convergence. We hope the algorithm will rapidly converge to the required least squares solution. Unfortunately, in some cases, the iteration may converge to a local minimum, or do not converge at all. Please refer to [4, 8] for more details on nonlinear optimization.

C.2.2 Newton Iteration

Newton iteration is a basic algorithm used to solve nonlinear least squares problem. Suppose at the ith iteration, the function is approximated by $f(\mathbf{x}) = f(\mathbf{x}_i) + \varepsilon_i$, and the func-

C

tion can be locally linearized at $\mathbf{x}_i$ as

$$f(\mathbf{x}_i + \Delta_i) = f(\mathbf{x}_i) + \mathbf{J}\Delta_i$$

where $\mathbf{J}$ is the Jacobian matrix. We seek to find the next point $\mathbf{x}_{i+1} = \mathbf{x}_i + \Delta_i$ such that $f(\mathbf{x}_{i+1})$ is closer to $f(\mathbf{x})$. From

$$f(\mathbf{x}) - f(\mathbf{x}_{i+1}) = f(\mathbf{x}) - f(\mathbf{x}_i) - \mathbf{J}\Delta_i = \varepsilon_i - \mathbf{J}\Delta_i$$

we know that it is equivalent to minimization of

$$\min_{\Delta_i} \|\varepsilon_i - \mathbf{J}\Delta_i\|$$

which is exactly the linear least squares problem (C.2). It can be solved by the normal equations

$$\mathbf{J}^T\mathbf{J}\Delta_i = \mathbf{J}^T\varepsilon_i \tag{C.19}$$

Thus the increment is given by $\Delta_i = (\mathbf{J}^T\mathbf{J})^{-1}\mathbf{J}^T\varepsilon_i = \mathbf{J}^+\varepsilon_i$, and the parameter vector can be updated according to

$$\mathbf{x}_{i+1} = \mathbf{x}_i + \mathbf{J}^+\varepsilon_i \tag{C.20}$$

where the Jacobian matrix is evaluated at $\mathbf{x}_i$ in each iteration.

$$\mathbf{J} = \left.\frac{\partial f(\mathbf{x})}{\partial \mathbf{x}}\right|_{\mathbf{x}_i}$$

C.2.3 Levenberg-Marquardt Algorithm

The Levenberg-Marquardt (LM) algorithm is the most widely used optimization algorithm for bundle adjustment. The algorithm is a variation of Newton iteration. In LM algorithm, the normal equations (C.19) are replaced by the augmented normal equations

$$(\mathbf{J}^T\mathbf{J} + \lambda\mathbf{I})\Delta_i = \mathbf{J}^T\varepsilon_i \tag{C.21}$$

where $\mathbf{I}$ is the identity matrix. The damping factor λ is initially set to some value, typically 0.001. The update rule is as follows: if the error goes down following an update Δ_i, the increment is accepted and λ is divided by a factor (usually by a factor of 10) before next iteration to reduce the influence of gradient descent. On the other hand, if Δ_i leads to an increased error, then λ is multiplied by the same factor and the update Δ_i is solved again from (C.21). The process is repeated until an acceptable update is found.

The above algorithm has a disadvantage. If the damping factor λ is large, the matrix $\mathbf{J}^T\mathbf{J}$ in (C.21) is not used at all. This leads to a large movement along the directions where the gradient is small. To address this problem, Marquardt [6] modified equation (C.21) to

$$\left(\mathbf{J}^T\mathbf{J} + \lambda\,\mathrm{diag}(\mathbf{J}^T\mathbf{J})\right)\Delta_i = \mathbf{J}^T\varepsilon_i \tag{C.22}$$

where the identity matrix is replaced by the diagonal of $\mathbf{J}^T\mathbf{J}$. Each step of the LM algorithm involves the solution of normal equation (C.21) or (C.22), which has a complexity of n^3. Thus, it is computational intensive for a problem with large number of parameters.

In bundle adjustment (C.17), the parameter space consists of two different sets. One set of parameters is related to camera parameters and the other set consists of space points. This leads to a sparse structure in the Jacobian matrix. Therefore, a sparse Levenberg-Marquardt algorithm was proposed to improve its efficiency. Please refer to [3, 5] for implementation details.

References

1. Faugeras, O.: Three-Dimensional Computer Vision. MIT Press, Cambridge (1993)
2. Golub, G.H., Van Loan, C.F.: Matrix Computations, 3rd edn. Johns Hopkins University Press, Baltimore (1996). ISBN 978-0-8018-5414-9
3. Hartley, R.I., Zisserman, A.: Multiple View Geometry in Computer Vision, 2nd edn. Cambridge University Press, Cambridge (2004). ISBN: 0521540518
4. Kelley, C.T.: Iterative Methods for Optimization. SIAM Frontiers in Applied Mathematics, vol. 18. SIAM, Philadelphia (1999)
5. Lourakis, M.I.A., Argyros, A.A.: SBA: A software package for generic sparse bundle adjustment. ACM Trans. Math. Softw. **36**(1), 1–30 (2009)
6. Marquardt, D.: An algorithm for least-squares estimation of nonlinear parameters. SIAM J. Appl. Math. **11**, 431–441 (1963)
7. Mundy, J.L., Zisserman, A.: Geometric Invariance in Computer Vision. MIT Press, Cambridge (1992)
8. Nocedal, J., Wright, S.J.: Numerical Optimization. Springer, New York (1999)
9. Press, W.H., Flannery, B.P., Teukolsky, S.A., Vetterling, W.T.: Numerical Recipes in C. Cambridge University Press, Cambridge (1988)
10. Triggs, B., McLauchlan, P., Hartley, R. Hartley, Fitzgibbon, A.: Bundle adjustment—A modern synthesis. In: Proc. of the International Workshop on Vision Algorithms, pp. 298–372. Springer, Berlin (1999)

G. Wang, Q.M.J. Wu, *Guide to Three Dimensional Structure and Motion Factorization*, Advances in Pattern Recognition,
DOI 10.1007/978-0-85729-046-5,

Glossary

Affine factorization A structure and motion factorization algorithm based on affine camera model. Depending on the type of object structure, it is classified as rigid factorization, nonrigid factorization, articulated factorization, etc.

Bundle adjustment Any refinement approach for visual reconstructions that aims to produce jointly optimal structure and camera estimates.

Disparity The position difference of a feature point in two images, which is denoted by a shift of an image point from one image to the other.

Duality principle The proposition in projective geometry appears in pairs. Given any proposition of the pair, a dual result can be immediately inferred by interchanging the parts played by the words "point" and "line".

Extended Cholesky decomposition An extension of Cholesky decomposition to efficiently factorize a positive semidefinite matrix into a product of a vertical extended upper-triangular matrix and its transpose.

Feature matching The process of establishing correspondences between the features of two or more images. Typical image features include corners, lines, edges, conics, etc.

Hadamard product Refers to element-by-element multiplication of two matrices with same dimensions.

Histogram A graphical display of tabulated frequencies. It is the graphical version of a table which shows what proportion of cases fall into every specified categories.

Incomplete tracking matrix A tracking matrix with missing point entries due to tracking failures.

Lateral rotation The orientation of a camera is denoted by roll-pitch-yaw angles, where the roll angle is referred as axial rotation, and the pitch and yaw angles are referred as lateral rotation.

Nonrigid factorization Structure and motion factorization algorithm for nonrigid objects. It may assume different camera models, such as affine, perspective projection model, or quasi-perspective projection model.

G. Wang, Q.M.J. Wu, *Guide to Three Dimensional Structure and Motion Factorization*, Advances in Pattern Recognition,
DOI 10.1007/978-0-85729-046-5, © Springer-Verlag London Limited 2011

Normalized tracking matrix Depending on the context, it refers either to a tracking matrix in which every point is normalized by the camera parameters, or a tracking matrix that is normalized point-wisely and image-wisely so as to increase its numerical stability and accuracy in factorization.

Perspective factorization Perspective projection-based structure and motion factorization algorithm for either rigid or nonrigid objects.

Pose estimation Estimating the position and orientation of an object relative to the camera, which are usually expressed by a rotation matrix and a translation vector.

Projective factorization Refers to structure and motion factorization in a projective space, which may be stratified to Euclidean space by an upgrading matrix.

Quasi-perspective factorization Structure and motion factorization of rigid or nonrigid objects based on quasi-perspective projection that assumes the camera is far away from the object and undergoes small lateral rotations. The projective depths are implicitly embedded in the motion and shape matrices in quasi-perspective factorization.

Rigid factorization Structure and motion factorization of rigid objects. Its formulation is depended on the employed projection model, such as affine camera model, perspective projection model, or quasi-perspective projection model.

Stratified reconstruction A 3D reconstruction approach which begins with a projective or affine reconstruction, then refines the solution progressively to an Euclidean solution by applying constraints about the scene and the camera.

Structure and motion factorization The process of factorizing a tracking matrix into 3D structure and motion parameters associated with each frame. It is usually carried out by SVD decomposition or power factorization algorithm.

Tracking matrix A matrix composed by coordinates of tracked points across an image sequence. It is also termed as a measurement matrix.

Visual metrology The science of acquiring measurement through computer vision technology.

Weighted tracking matrix A tracking matrix in which the coordinates of every tracked point is scaled by a projective depth. In presence of a set of consistent depth scales, the factorization produces a perspective solution.

Index

G. Wang, Q.M.J. Wu, *Guide to Three Dimensional Structure and Motion Factorization*, Advances in Pattern Recognition,
DOI 10.1007/978-0-85729-046-5,

MIX
Papier aus verantwortungsvollen Quellen
Paper from responsible sources
FSC® C105338

If you have any concerns about our products, you can contact us on
ProductSafety@springernature.com

In case Publisher is established outside the EU, the EU authorized representative is:
Springer Nature Customer Service Center GmbH
Europaplatz 3, 69115 Heidelberg, Germany

Printed by Libri Plureos GmbH
in Hamburg, Germany